90 0568204 5

WITHDRAWN
FROM
UNIVERSITY OF PLYMOUTH
LIBRARY SERVICES

D1680979

This book is to be returned on
or before the date stamped below

I.L.L.

2 7 FEB 2008

UNIVERSITY OF PLYMOUTH

PLYMOUTH LIBRARY

Tel: (01752) 232323
This book is subject to recall if required by another reader
Books may be renewed by phone
CHARGES WILL BE MADE FOR OVERDUE BOOKS

Urtica

Medicinal and aromatic plants – industrial profiles

Individual volumes in this series provide both industry and academia with in-depth coverage of one major genus of industrial importance.
Edited by Dr Roland Hardman

Urtica

Therapeutic and nutritional aspects of stinging nettles

Edited by

Gulsel M. Kavalali

University of Istanbul, Turkey

Taylor & Francis
Taylor & Francis Group

LONDON AND NEW YORK

First published 2003
by Taylor & Francis
11 New Fetter Lane, London EC4P 4EE

Simultaneously published in the USA and Canada
by Taylor & Francis Inc,
29 West 35th Street, New York, NY 10001

Taylor & Francis is an imprint of the Taylor & Francis Group

© 2003 Taylor & Francis Ltd

Typeset in Garamond by
Newgen Imaging Systems (P) Ltd, Chennai, India
Printed and bound in Great Britain by
TJ International Ltd, Padstow, Cornwall

All rights reserved. No part of this book may be reprinted or reproduced
or utilised in any form or by any electronic, mechanical, or other means,
now known or hereafter invented, including photocopying and recording,
or in any information storage or retrieval system, without permission in
writing from the publishers.

Every effort has been made to ensure that the advice and information in this book
is true and accurate at the time of going to press. However, neither the publisher
nor the authors can accept any legal responsibility or liability for any errors or
omissions that may be made. In the case of drug administration, any medical
procedure or the use of technical equipment mentioned within this book,
you are strongly advised to consult the manufacturer's guidelines.

British Library Cataloguing in Publication Data
A catalogue record for this book is available from the British Library

Library of Congress Cataloging in Publication Data
 Urtica: therapeutic and nutritional aspects of stinging nettles /
edited by Gulsel Kavalali-
  p. cm. (Medicinal and aromatic plants—industrial
profiles ; v. 37)
  Includes bibliographical references and index.
  1. Nettles–Therapeutic use. I. Kavalali, Gulsel. II. Series.
[DNLM: 1. Urticaceae. 2. Phytotherapy. QV 766 U82 2003]
RM666 .N44U78 2003
615'.32345–dc21   2003005513

ISBN 0–415–30833–X

Dedicated to the memory of my father and mother

UNIVERSITY OF PLYMOUTH		
Item No.	9005682045	
Date	17 SEP 2003 S	
Class No.	615.32345 URT	
Cont. No.	✓	
PLYMOUTH LIBRARY		

Contents

Tables

Contributors

Gulsel Kavalali, Herbal Medicine Research and Development Center, University of Istanbul, Cerrahpasa Faculty of Medicine, Cerrahpasa 34303, Istanbul, Turkey.

Johannes Josef Lichius, Institut für Pharmazeutische Biologie und Pharmakologie, Martin-Luther-Universität Halle-Wittenberg, Germany.

Colin Randall, Departments of General Practice and Primary Care, Peninsula Medical School, and Rheumatology, Derriford Hospital, Plymouth, UK.

Huriye Wetherilt, Lalecan Nutrition and Food Consultancy Ltd, Bagdat Cad., Selvili Sokak, Selvi Apt. 10/3, Suadiye, Istanbul, Turkey.

Preface to the series

There is increasing interest in industry, academia and the health sciences in medicinal and aromatic plants. In passing from plant production to the eventual product used by the public, many sciences are involved. This series brings together information which is currently scattered through an ever increasing number of journals. Each volume gives an in-depth look at one plant genus, about which an area specialist has assembled information ranging from the production of the plant to market trends and quality control.

Many industries are involved, such as forestry, agriculture, chemical, food, flavour, beverage, pharmaceutical, cosmetic and fragrance. The plant raw materials are roots, rhizomes, bulbs, leaves, stems, barks, wood, flowers, fruits and seeds. These yield gums, resins, essential (volatile) oils, fixed oils, waxes, juices, extracts and spices for medicinal and aromatic purposes. All these commodities are traded worldwide. A dealer's market report for an item may say 'Drought in the country of origin has forced up prices'.

Natural products do not mean safe products and an account of this has to be taken by the above industries, which are subject to regulation. For example, a number of plants which are approved for use in medicine must not be used in cosmetic products.

The assessment of safe-to-use starts with the harvested plant material which has to comply with an official monograph. This may require absence of, or prescribed limits of, radioactive material, heavy metals, aflatoxin, pesticide residue, as well as the required level of active principle. This analytical control is costly and tends to exclude small batches of plant material. Large scale contracted mechanized cultivation with designated seed or plantlets is now preferable.

Today, plant selection is not only for the yield of active principle, but for the plant's ability to overcome disease, climatic stress and the hazards caused by mankind. Such methods as *in vitro* fertilization, meristem cultures and somatic embryogenesis are used. The transfer of sections of DNA is giving rise to controversy in the case of some end-uses of the plant material.

Some suppliers of plant raw material are now able to certify that they are supplying organically-farmed medicinal plants, herbs and spices. The Economic Union directive (CVO/EU No 2092/91) details the specifications for the *obligatory* quality controls to be carried out at all stages of production and processing of organic products.

Fascinating plant folklore and ethnopharmacology leads to medicinal potential. Examples are the muscle relaxants based on the arrow poison, curare, from species of *Chondrodendron*, and the anti-malarials derived from species of *Cinchona* and *Artemisia*.

The methods of detection of pharmacological activity have become increasingly reliable and specific, frequently involving enzymes in bioassays and avoiding the use of laboratory animals. By using bioassay-linked fractionation of crude plant juices or extracts, compounds can be specifically targeted which, for example, inhibit blood platelet aggregation, or have anti-tumour, or anti-viral, or any other required activity. With the assistance of robotic devices, all the members of a genus may be readily screened. However, the plant material must be *fully* authenticated by a specialist.

The medicinal traditions of ancient civilisations such as those of China and India have a large armamentaria of plants in their pharmacopoeias which are used throughout South-East Asia. A similar situation exists in Africa and South America. Thus, a very high percentage of the World's population relies on medicinal and aromatic plants for their medicine. Western medicine is also responding. Already in Germany all medical practitioners have to pass an examination in phytotherapy before being allowed to practise. It is noticeable that throughout Europe and the USA, medical, pharmacy and health related schools are increasingly offering training in phytotherapy.

Multinational pharmaceutical companies have become less enamoured of the single compound magic bullet cure. The high costs of such ventures and the endless competition from 'me too' compounds from rival companies often discourage the attempt. Independent phyto-medicine companies have been very strong in Germany. However, by the end of 1995, eleven (almost all) had been acquired by the multinational pharmaceutical firms, acknowledging the lay public's growing demand for phytomedicines in the Western World.

The business of dietary supplements in the Western World has expanded from the health store to the pharmacy. Alternative medicine includes plant-based products. Appropriate measures to ensure the quality, safety and efficacy of these either already exist or are being answered by greater legislative control by such bodies as the Food and Drug Administration of the USA and the recently created European Agency for the Evaluation of Medicinal Products, based in London.

In the USA, the Dietary Supplement and Health Education Act of 1994 recognized the class of phytotherapeutic agents derived from medicinal and aromatic plants. Furthermore, under public pressure, the US Congress set up an Office of Alternative Medicine and this office in 1994 assisted the filing of several Investigational New Drug (IND) applications, required for clinical trials of some Chinese herbal preparations. The significance of these applications was that each Chinese preparation involved several plants and yet was handled as a *single* IND. A demonstration of the contribution to efficacy, of *each* ingredient of *each* plant, was not required. This was a major step forward towards more sensible regulations in regard to phytomedicines.

My thanks are due to the staff of Taylor & Francis publishers who have made this series possible and especially to the volume editors and their chapter contributors for the authoritative information.

Roland Hardman, 1997

Preface

I grew up in a small country town in the west of Anatolia, Turkey. During my childhood, I was acquainted with nature through my father who was a biology teacher in a high school. Together we collected plant and stone specimens.

After graduating as a pharmacist and acquiring a Ph.D degree in pharmacognosy, I elucidated chemical compounds from different plants, some of which had attractive flowers and good odors. Later, I joined the pharmacology department in the medical school where I carried out studies on *Urtica*, which was a terrible plant, that had no attractive flowers, no pleasant odor and which grew in desolate areas. It was difficult to collect the specimens as they burned and hurt me. However, after I began to study and elucidate some active compounds from *Urtica,* I learned to love this plant and thought it would be useful and beneficial for human life in the future.

Over the years, scientific researches have expanded and made more precise our knowledge of the chemical effects and composition of the active constituents which determine the medicinal properties of plants. Modern pharmaceutical industry makes use of medicinal plants, especially in the dried form as crude drugs and as raw material in the manufacture of medicinal preparations.

In the first half of the nineteenth century apothecaries not only stocked dried medicinal plants in the form of crude drugs for the preparation of various herbal tea mixtures, but also used them to make all kinds of tinctures, extracts and juices which in turn were employed in preparing medicinal drops, syrups, infusions, ointments and liniments. Certain medicinal plants are still in use in the form of teas. During the second half of the nineteenth century, several important discoveries in the developing field of chemistry enabled this science to progress rapidly.

Medicinal plants became one of chemistry's chief objects of interest, and in time chemists succeeded in isolating the pure active substances. Groups of substances replaced the crude drugs. Then came the first synthetic medicines which became predominant and superseded the herbal medicines.

Medicinal plants are the basic material for making up herbal teas, taken either in the form of decoctions or as infusions. Medicinal plants are also used in industrial food products especially in dietetic preparations of aromatic substances, vitamins, important amino acids and enzymes.

Medicinal plants and the investigation of their chemical composition, as well as of their therapeutic effects, yield the knowledge used in the synthetic preparation of new substances and provide substances, which in various combinations become powerful medicinal agents.

Presently, medicinal plants are important in the form of crude drugs and for the medically active materials isolated from them. This is an important aid to the physicians in the treatment of diseases.

I believe this book will be of use to physicians and pharmacists as a source of reference to further studies which develop new drugs in the pharmaceutical industry. Additionally, it will be useful to the practitioners of complementary drug therapy.

Acknowledgements

First of all, I would like to thank Dr R. Hardman for encouraging me to write this book and also Professor A. Baytop for providing helpful information on botanical aspects of *Urtica*.

I have been fortunate to have had the access to library and literature services of several universities and pharmaceutical companies.

The preparation of the manuscripts would not have been possible without aid from several sources; thanks to Dr B. Mathew for arranging my visit to the Royal Botanic Gardens (Kew) and also many thanks to Dr M. Thomas and C. Barker for their assistance when I studied *Urtica* specimens at the Kew Herbarium.

I express my gratitude to Professor W. J. Peumans and Professor H. Wagner. They both kindly sent me their original papers and allowed me to use them in this book. Also, special thanks to Dr J. H. Zwaving for providing the Escop Monograph and some additional information.

I deeply appreciate the contributors' valuable cooperation and their timely completion of works. Also, special thanks to Dr H. Wetherilt for checking the manuscript and offering helpful suggestions.

Gulsel M. Kavalali
Istanbul, December 2002

1　An introduction to *Urtica* (botanical aspects)

Gulsel Kavalali

Introduction

Urtica L., the stinging nettle (Urticaceae), is an annual and perennial herb, distinguished with stinging hairs. The leaves are opposite, and the flowers are green with yellow stamens. The male and female flowers are on separate plants. The fruits are achene. These are the characteristics of the *Urtica* genus which belong to the family Urticaceae. The main varieties identified under the *Urtica* species are *U. dioica* L., *U. urens* L., *U. pilulifera* L., *U. cannabina* L., *U. membranacea* Poiret., *U. haussknechtii* Boiss., *U. atrovirens* Req., *U. rupestris* Guss., *U. chamaedryoides* Pursh., *U. ferox* Forst.

Among these, *U. dioica* and *U. urens* have been known for a long time as medicinal plants. They are used as an expectorant, purgative, diuretic, haemostatic, vermifuge and for the treatment of eczema, rheumatism, haemorrhoids, hyperthyroidism, bronchitis and cancer. These plants have been consumed without any report of serious adverse effects. They are also healthy vegetables and potential organic fertilizer for herbs. The commercially available drug preparations known as *Urtica* folium/herb and *Urtica* radix – which are obtained from the *U. dioica* and *U. urens* – are produced in Europe.

The generic name 'Urtica' comes from the latin word 'Uro' meaning 'to burn'. 'Nettle' comes from the Anglo-Saxon word 'Netele' referring to the sharp stings or the use of the fibrous nettle stems in cloth-making.

Botanical aspects of *Urtica*

Description of genus

The genus *Urtica* L. belongs to the family Urticaceae, under the division Spermatophyta, subdivision of Angiospermae, class Dicotyledonae, group Apetalae, order Urticales.

The family Urticaceae

The family Urticaceae comprises some 40 genera and more than 700 species of monoecious or dioecious plants which mostly grow in tropical and subtropical regions of the world. The plants are generally fibrous herbs, rarely subshrubs, softwooded small trees or climbers. Some have stinging hairs on their stems and leaves. These are silicified epidermal outgrowths containing an irritating cell sap. Latex is absent, cystoliths of various shapes are present in the epidermal cell. The leaves are alternate or opposite, usually stipulate. Inflorescences are axiliary, basically a bracteate cyme, sometimes a head or reduced to a single flower. The flowers are unisexual, small, mostly green, actinomorphic, (tri-) tetra- and pentamerous. The perianth is of (three) four to five distinct or connate segments. The male flowers have (one) four to five stamens; the filaments are bent inwards in the bud and spring back elastically at the anthesis, releasing pollen in a sudden burst. The female flowers have a single pistil, an ovary unilocular with a solitary ovule, and a stigma which is often brushlike. The fruit is a dry achene or more rarely a drupe, which is often enclosed by the persistent perianth. The seeds have an oily endosperm and a straight embryo (Lawrence, 1951; Hess *et al.*, 1967; Chaurasia, 1987). Important characteristics of the family including: mostly herbaceous habit; usually stipulate leaves, often with cystoliths and stinging hairs; unisexual and monochlamydeous small flowers with inwardly bent stamens, a brushlike stigma, and a solitary basal orthotropous ovule.

Five tribes are recognized in this family (Hegnauer, 1973): Urticeae, Procrideae, Boehmerieae, Parietarieae, Forskohleae. Stinging hairs are only present in Urticeae. Economically the family is important because of the fibre yielded by some of the members. The most important is Ramie (*Boehmeria nivea* var. *tenacissima*), native of East Asia, also cultivated today in other tropical and subtropical regions of the world. The fibre of the *U. dioica* is also important, because of its high tensile strength and non-lignified walls. Another importance of the family comes from the leaves of the *Urtica* and *Parietaria* species which are used in folk medicine. Furthermore, young *Urtica* leaves serve as an edible green. As ornamental plants, species of *Pilea* and *Pellonia* are grown under glass. *Soleirolia* is cultivated in shady gardens for covering the ground.

The genus Urtica L. Linn. Syst. ed. I (1735)

These are annual or perennial herbs but a few are shrubs and small trees, furnished with stinging hairs. Stinging hairs are stiff, long, unicellular and the apex bears a small spherical or ovoid head which is attached obliquely and readily breaks off upon contact, the irritating cell fluid thus discharged penetrates into the skin. The leaves are opposite, dentate, stipulate, stipules free or connate in pairs. Inflorescence axillary in false spikes, sometimes condensed in globose heads. The flowers are unisexual, small, green. The male flowers have four equal perianth segments and four stamens inwardly bent in bud, springing back at anthesis. The female flowers have four perianth segments in two pairs, the inner pair is accrescent in fruit and encloses the small achene (Ball, 1964; Townsend, 1982).

The main varieties identified under the *Urtica* species are listed as follows:

1   *Urtica dioica* L.   Sp. Pl. 984 (1753) (common nettle)
    It is a perennial herb, 30–150 or (250) cm in height, dull green in colour, usually dioecious. Leaves are ovate (4–11 × 3–10 cm) rarely lanceolate acuminate. The

Figure 1.1 Urtica pilulifera L. (roman nettle).

male and female inflorescences are similar in form, much branched. The female flowers have purplish stigma.

   This species is variable and contains a number of subspecies growing throughout the temperate regions of Europe and Asia.

2   *Urtica urens* L.   Sp. Pl. 984 (1753) (lesser nettle)
   It is an annual herb, monoecious, 10–60 cm in height, clear green in colour. Leaves are (1–4, 1–6 cm), ovate and deeply serrate. The male and female flowers are numerous, centrally glabrous or sparsely hispid on the back. It resembles the common nettle in habit but has smaller leaves and short flowers. It grows throughout Europe except the north.

3   *Urtica pilulifera* L.   Sp. Pl. 983 (1753) (roman nettle)
   It is an annual herb, monoecious, 30–100 cm in height. Leaves are ovate (2–6 cm) unisexual. Female flowers are in globose heads, the male flowers are in dense clusters. It is most often found in a Mediterranean climate (Figure 1.1).

4   *Urtica membranacea* Poiret in Lam.   Encycl. 4 : 638 (1798)
   It is an annual herb, monoecious, 15–80 cm in height. Leaves are ovate or triangular (8×7.5 cm) coarsely dentate. Petiole almost as long as the lamina. It grows in southern Europe, northern Africa and south-west Asia.

   This species is usually described as monoecious but its sexuality seems variable (Davis, 1982).

5   *Urtica haussknechtii* Boiss   Fl. or 4 : 1146 (1875)
   It is a perennial herb with many spreading branches, but only young branchlets

have stinging hairs. Leaves are lanceolate, largest can grow to 5×1.4 cm; stinging hairs are not apparent.

According to Davis (1982), this species can be a typical secondary growth of *U. dioica* but the leaves are not in narrow-leaved form.

6   *Urtica atrovirens* Req. ex Loisel.   Mém. Soc Linn. Paris 6 : 432 (1827)

It is a perennial herb, monoecious, 30–100 cm in height, dull green in colour. Leaves are lanceolate or ovate (2×7 cm). Racemes are longer than petiole with both male and female flowers. Female flowers are pubescent. It grows in the western Mediterranean region.

7   *Urtica cannabina* L. Sp. Pl. 984 (1753)

This species is typically *U. dioica* but it is 200 cm in height with stinging hairs. It grows in Russia.

8   *Urtica rupestris* Guss. Cat. Pl. Boccad. 83 (1821)

It is a perennial herb, dioecious, 30–100 cm in height. Leaves are 2×10 cm ovate-acuminate. Racemes are unisexual. The female is not more than 2 cm but the male is 2–8 cm. It grows in Sicilia.

9   *Urtica chamaedryoides* Pursh.   Fl. Am. Sept. i.118.

It is an annual herb, 20–70 cm in height. Stinging bristles, leaves are ovate and mostly cordate. It grows in the United States.

10   *Urtica ferox* Forst. f.   Prod. 66.-

This species is a tree, 3 m in height, known as ongaonga. It grows in New Zealand and North-South Island.

Distribution of *Urtica*

The genus comprises *c.*100 species, distributed in temperate and tropical regions throughout the world. But some *Urtica* species are grown as food plants for a nutrition element. According to the Index Kewensis (1895–1995), the registered species of *Urtica* are listed in the Table 1.1. Some other registered *Urtica* species are listed in the Tables 1.2–1.9.

Table 1.1 The species of *Urtica* registered in Index Kewensis (1895–1995)

Species name	Distribution
Urtica	
africana Linn. ex Jackson	Quid
arborescens Link	Ins. Philipp.
aspera Petric.	N. Zel.
atrovirens Req. ex Loisel.	Europ. Austr.
aucklandica Hook. f.	Ins. Aucland
australis Hook.f.	N. Zel.
baccifera Sessé et Moc.	Mexico
ballotaefolia Wedd.	N. Granat.
biserrata Sessé et Moc.	Mexico

(Continued)

Table 1.1 (Continued)

Species name	Distribution
Breweri S.Wats.	Am. Bor. occ.
Buchtienii Ross	Chili
Buraei Léveillé et Vaniot	China
Burchellii N. E. Brown	Afr. Austr. (Cape Prov.)
Californica Greene	Calif.
calophylla Schlecht.	Hab.?
cannabina L.	Asiabor., Persia
caracasana Griseb.	Reg. Argent.
cardiophylla Rydb.	Am. Bor.
chamaedryoides Pursh.	Am. Bor.
chichicaztli Sessé et Moc.	Mexico
cinerascens Poir.	Ind. or
contracta Blume	Java
copeyana Killip	Costa Rica
cordatifolia Steud.	China
crenulata Roxb.	Bengal
cubensis Klotzsch ex Herd.	Cuba
cyanescens Komarov	Sibir.
dentata Hand.-Mazz.	China (Hunan)
dioica Linn.	Reg. Bor. temp.
domingensis Urb.	Haiti
echinata Benth.	Am. Austr.
elongata Link	Ins. Philipp.
elongata Vahl	Ind. occ.
fastigiata Blume	Chili
ferox Forst. f.	N. Zel.
Finlaysoniana Wall.	Siam?
Fischeri Engl.	Afr. trop. or
fissa E. Pritz. ex Diels	China
flabellata H. B. et K.	Ecuador
foliosa Blume	Japan
fragilis J. Thiéb.	Libanus
frutescens Poir.	Am. Austr.
galeopsifolia Jacq. f. ex Blume	Hab.?
glabrata Clem. ex Vis.	Locality not stated
glandulifera Liebm.	Mexico
glanbulifera Roxb.	Ind. or
glomeruliflora Steud.	Ins. J. Fernand
gracilenta Greene	Mexico
gradidentata Miq.	Java
granulosa S. F. Blake	Mexico
Haussknechtii Boiss.	Cataonia
holosericea Nutt.	Am. Bor. occ.
hulensis Feinbr.	Palaest.
hyberborea Jacquem. ex Wedd.	Reg. Himal

(Continued)

Table 1.1 (Continued)

Species name	Distribution
Ildefonsia Steud.	Bras.
incisa Poir.	Austral., N. Zel.
inermis Spreng.	Ins. Pacif.
intermedia Form.	Europe
kunlunshanica C. Y. Yang	Xinjiang
laetevirens Maxim.	Japan
latifolia Moon	Zeylan
laurifolia Poir.	Reg. Magellan
lineata Roxb.	Ins. Molucc.
lobatifolia Ying	Taiwan
longifolia Sessé et Moc.	Mexico
Lyallii S. Wats.	Am. Bor. occ.
Macbridei Killip	Peruv.
macrorrhiza Hand.-Mazz.	China (Yunnan)
magellanica Juss. ex Poir.	Reg. Magellan
Mairei Léveillé	China
Masafuerae Phil.	Ins. Masafuer
massaica Mildbr.	Tanganyika Terr.
membranacea Poir. ex Savigny	Reg. Mediterr.
membranifolia C. J. Chen	China (Xizang, Sikkim)
mexicana Liebm.	Mexico
Meyeri Wedd.	Afr. Austr.
micrantha Kunth et Bouché	Hab.?
Millettii Hook. et Arn.	China
minima Roxb.	Ins. Mollucc.
minutifolia Griseb	Reg. Argent.
mollis Steud.	Chili
moluccana Blume	Ins. Mollucc.
morifolia Poir.	Ins. Canar.
muralis Vahl	Arab.
nana Hook.	N. Granat.
neubaueri Chrtek	Afganistan
nicaraguensis Liebm.	Nicaragua
nudula Vogel	Ind. occ. (S. Kitts)
ovalifolia Blume	Java
pachyrrhachis Hand.-Mazz.	China (Hunan)
papuana Zandee	N. Guin.
parviflora Roxb.	Reg. Himal
peruviana Geltman	Peru
pilulifera L.	Reg. Mediterr., Oriens, Ind. or
pinfaensis Lévéillé et Blin	China (Kweichau)
pulchella Hort. Angl. ex Link	Ind. or
pseudomagellanica Geltman	Bolivia
rotundifolia Sessé et Moc.	Portoric.
rubricaulis Herd. ex Regel	Hab.?
rubricoulis Span.	Ins. Timor

(Continued)

Table 1.1 (Continued)

Species name	Distribution
rupestris Guss.	Sicil.
sansibarica Engl.	Afr. trop. or
Serra Blume	Mexico
sikokiana Makino	Japan
silvatica Hand.-Mazz.	China
simensis Hoch. ex A Rich.	Ethiopia
spatulata Sm.	Am. Austr.
spicata Sessé et Moc.	Mexico
spirealis Blume	Mexico
stachyoides Webb et Berth	Ins. Canar
stipulacea Bertol	As. Min.
strigosissima Rydberg	USA (Idaho)
subincisa Benth	Mexico
taiwaniana S. S. Ying	Taiwan
Thunbergiana Sieb.et Zucc.	Japan
tibetica W.T.Wang ex C. J. Chen	China (Xizang)
Tremolsii Sennen	Hispan
Trianae Rusby	Boliv.
triangularis Hand.-Mazz.	China (Szechuan, Yunnan)
urens L.	Geront. temp
urescens Rojans	Argent.
villosa Thunb.	Japan
viridis Rydberg	Montana. Wyoming, Idaho, Alberta
Wallichiana Steud.	Ins. Molucc.
whitlowii Whitlow	Locality not stated
xiphodon Stapf	As. occ.
zayuensis C. J. Chen	China (Xizang)

Table 1.2 The species of *Urtica* recorded in Flora URSS

Species name	Distribution
Urtica	
angustifolia Fisch. ex Hornem	E. Siberia, Far East
cannabina L.	W. and E. Siberia, C. Asia, Far East
cyanescens Kom.	Far East
dioica L.	Caucasia, W. and E. Siberia, C. Asia, Far East
kioviensis Rogow.	Kiev, Upper and Middle Dnieper
laetevirens Maxim.	Far East
pilulifera L.	Caucasia
platyphylla Wedd.	Far East
pubescens Ledeb.	Don, Caucasia
urens L.	Caucasia, W. and E. Siberia, Far East

Source: Komarov (1936).

Table 1.3 The species of *Urtica* which occur in Europe

Species name	Distribution
Urtica	
atrovirens Req. ex Loisel.	W. Mediterranean region
cannabina L.	Naturalized in Russia
dioica L.	Throughout Europe
dubia Forskal	Mediterranean region, Portugal, Açores
kioviensis Rogow.	E. C. Europe, from E. Austria to C. Ukraine
pilulifera L.	S. Europe, often naturalized or casual elsewhere
rupestris Guss.	Sicilia
urens L.	Throughout Europe except the extreme North

Source: Ball (1964).

Table 1.4 The species of *Urtica* which occur in Turkey

Species name	Distribution
Urtica	
dioica L.	Widespread in Turkey-in-Europe and Anatolia
Haussknechtii Boiss.	Known only in old Malatya
membranacea Poiret	W. Turkey, S. Anatolia
pilulifera L.	Turkey-in-Europe, outer and adjacent C. Anatolia
urens L.	Scattered in Anatolia

Source: Townsend (1982).

Table 1.5 The species of *Urtica* which occur in Iran

Species name	Distribution
Urtica	
cannabina L.	N. Persia
dioica L.	Throughout Persia
pilulifera L.	N., W. and S. Persia
urens L.	N., W. and S. Persia

Source: Chrtek (1974).

Table 1.6 The species of *Urtica* which occur in Iraq

Species name
Urtica
dioica L.
pilulifera L.
urens L.

Source: Townsend (1980).

Table 1.7 The species of *Urtica* which occur in Syria, Lebanon and Palestine

Species name	Distribution
Urtica	
dioica L.	Syria, Lebanon, Palestine
fragilis Thiébaut	Syria, Lebanon
hulensis Feinbr.	Syria, Lebanon
membranacea Poiret	Lebanon, Palestine
pilulifera L.	Syria, Lebanon, Palestine
urens L.	Syria, Lebanon, Palestine

Source: Post and Dinsmore (1933), Mouterde (1966), Zohary (1966).

Table 1.8 The species of *Urtica* which occur in N. America

Species name
Urtica
chamaedryoides Pursh
dioica L.
gracilis Ait.
procera L.
urens L.
viridis Rydb

Source: Fernald (1950).

Table 1.9 The species of *Urtica* which
         occur in S. Africa

Species name
Urtica
dioica L.
lobulata E. Mey.
urens L.

Source: Watt and Breyer-Brandwijk (1962).

Regarding general distribution, the most common *Urtica* species seems to be the perennial *U. dioica*, followed by the annual *U. urens, U. pilulifera* and *U. membranacea*. *U. dioica* shows a wide range of distribution extending from Europe, N. Africa, to Asia, up to North and South America, and South Africa. *U. urens* occurs in Europe, N. Africa, Asia, N. America and S. Africa. *U. pilulifera* is more restricted in distribution. It is found in W. and S. Europe, N. Africa and S. W. Asia. *U. membranacea* is a Mediterranean element and occurs in S. Europe, N. Africa and S.W. Asia.

References

Ball, P. W. (1964) *Urtica* L. in T. G. Tutin *et al.* (eds) *Flora Europea*, Vol. 1, pp. 67–68, University Press, Cambridge.

Chaurasia, N. (1987) *Phytochemische Untersuchungen Flora Urtica dioicae* L. Inaugural-Dissertation zur Erlangung der Doktorwürde des Fachbereichs Pharmazie und Lebensmittelchemie der Philipps-Universitat Marburg/Lahn.

Chrtek, J. (1974) Urticaceae, in K. H. Rechinger (ed.) *Flora Iranica*, Lfg 105, 2–6, Akademische Druck-u. Verlagsanstalt, Graz.

Davis, P. H. (1982) *Flora of Turkey and The East Aegean Islands*, Vol. 7, p. 633, The University Press, Edinburg.

Fernald, L. M. (1950) *Gray's Manual of Botany*, eighth edn, pp. 556–559, American Book Company.

Hegnauer, R. (1973) *Chemataxonomy der Pflanzen*, Bd 6, pp. 629–637, Birkhäuser Verlag, Basel und Stuttgart.

Hess, H. E., Landolt, E. Hirzel, R. (1967) *Flora der Schweiz*, Bd 1, pp. 705–708, Birkhäuser Verlag, Basel und Stutugart.

Index Kewensis, (1895–1995) Vol. I, Vol. II and Suppl. 1–20 Clarendon Press, Oxford published by Royal Botanic Gardens (London, UK) publication date January 1997.

Komarov V. L. (ed.) (1936) Flora URSS, 5, pp. 386–394 Editio Academiae Scientiarum URSS, Mosqua.

Lawrence, G. H. M. (1951) *Taxonomy of Vascular Plants*, pp. 464–465, The Macmillan Company, New York.

Mouterde, P. (1966) *Nouvella Flore du Liban et de la Syrie*, tome premier, Texte, pp. 372–377, Imprimerie Catholique, Beyrouth.

Post, D. E. and Dinsmore, J. E. (1933) *Flora of Syria, Palestina and Sinai*, second edn, Vol. 2, pp. 510–512, American Press, Beirut.

Townsend, C. C. (1980) Urticaceae, in C. C. Townsend and E. Guest (eds), *Flora of Iraq*, Vol. 4, pp. 93–105, Robert MacLehose and Cie, Baghdad.

Townsend, C. C. (1982) Urticaceae, in P. H. Davis (ed.), *Flora of Turkey and the East Aegean Islands*, Vol. 7, pp. 633–638, University Press, Edinburgh.

Watt, J. M. and Breyer-Brandwijk, M. G. (1962) *The Medicinal and Poisonous Plants of Southern and Eastern Africa*. (1042–1046) E. S. Livingstone Ltd., Edinburgh and London.

Zohary, M. (1966) *Flora Palaestina*, Part one, Text, pp. 39–43, Goldberg's Press. Jerusalem.

2 Historical and modern uses of *Urtica*

Colin Randall

Introduction

In this chapter, the general historical background of the use of *Urtica* (nettles) is reported. Records of the use of nettles go back at least to the Bronze Age (3000–2000 BCE). The various ways in which nettles are currently used are also presented.

Non-medical uses of Urtica

Food

Nettles have traditionally been used as a nutritious food, particularly in the spring time in rural areas. This use has been widely reported in the United Kingdom (Grieve, 1983; Vickery, 1993; Beith, 1995; Mabey, 1996) and by the Thompson Indians of British Columbia in Canada (Turner *et al.*, 1990). It is used as a pot plant (nettle kail), nettle soup, nettle tea, and nettle ale (Vickery, 1995; Beith, 1995). There are reports of the Romans eating nettles as food and the recipe of St Columba's broth (St Columba was the sixth century Irish monk and poet) survives to this day:

> Pick young stinging nettles before the end of June, when they are four or five inches high – one handful for each person. Boil, drain, chop and return to pan with water and milk. Reheat, sprinkle in fine oatmeal or oats, stirring until thick. For present-day tastes eat with toast and grated cheese, or peeled soft boiled egg.
>
> (Mabey, 1996)

In Greek and Roman times nettle root was used in the boiling of meat to tenderize it (Bostock and Riley, 1890). Vickery records the perceived health-giving food value of nettles in Scotland and Ireland:

> In 1884 it was recorded that: nettles in many parts of Scotland were till not very many years ago used as a food, and were looked upon as a wholesome diet. The young and tender leaves were gathered, boiled, and then mashed ... mixed with a little oat-meal, and re-boiled for a short time. They were cooked in the same way as 'greens', which were and are still thought to possess medical virtues.

Particularly in Ireland, the use of nettles as a health-giving food continues to be recorded: then there was nettles, we had to gather them to make soup, oh, it was lovely … It used to be the custom to cook fresh young nettles and eat them as a kind of pick-me-up in the late spring, as they are rich in iron.

(Vickery, 1993)

In Sweden nettle tops are used as a tonic in the spring or in convalescence after illness (Robertson, 1998). Present day cooks still use the nettle and the famous chef, Anton Mosimann, blends his nettles with fromage blanc, new potatoes and nutmeg into a 'nettle nouvelle' (Mabey, 1996). Grieve recounts the folklore use of nettle juice with strong saline as vegetarian rennet in cheese making (Grieve, 1992). In Cornwall, UK, Yarg cheese is produced using nettle vegetarian rennet and the finished cheeses are wrapped in nettle leaves to preserve them (Randall, 2002). Nettles were also a prized food for turkeys in rural Britain in the 1930–1940 period (Vickery, 1993).

Mary Grieve also mentions the supposed benefits for animals: the claimed increased milk production of cattle fed on wilted nettles (in Sweden and Russia), increased turkey egg production, and a stomach remedy for horses. Nettle seeds added to feed are said to enhance a sleek coat in horses (in Holland and Egypt) (Grieve, 1992). Mills also recounts that the inclusion of nettles in the diet of breast-feeding humans increases milk supply (Mills, 1988).

Fibre for fabrics, threads, and paper

The use of nettle fibre to produce fabrics can be traced back to Bronze Age sites in Denmark where nettle fabric burial shrouds have been discovered (Patten, 1993). In Germany during the First World War (1914–1918) nettle cloth was manufactured when cotton was in short supply (Castleman, 1991). Nettle fibre has been used for sailcloth, sacking, and cordage. The fine cloth produced is similar in feel and appearance to silky linen (Ahket, 1983). Nettles are used by the Thompson Indians in British Columbia to produce thread for fishing nets (Turner *et al.*, 1990). In Germany in 1915 a very large area was given over to the commercial cropping of nettles to produce fabric, but the process of separation of fibres was difficult. Other fibres such as flax were cheaper to produce and nettle fibre production lost popularity (Grieve, 1983). Nettle fibre has also been traditionally used to manufacture paper (Grieve, 1983).

Dyes

A decoction of nettle plant produces a green dye, which has been used in Russia to dye woollen materials (Grieve, 1983). At the beginning of the Second World War nettle was one of the plants investigated by the British government for potential use. A request was made for the collection of 100 tons of nettles, and these were used for the extraction of a green dye for camouflage (Vickery, 1993).

Various other uses

Other uses reported for nettle are as a fly repellent (Ahket, 1983), as an additive to compost as an invigorator (Castleman, 1991), and as a spray to deter cabbage white butterflies from laying their eggs on cabbage crops (personal communication). Grieve reports the traditional use of nettle juice to seal leaks in wooden barrels. This and its traditional use as an anti-haemorrhaging agent, and a vegetarian rennet for cheese making are perhaps related to its agglutinating properties (Grieve, 1992). UDA – *Urtica dioica* (super) agglutinin – has recently been discovered and would seem to explain these examples of its uses (Peumans *et al.*, 1984; Shibuya *et al.*, 1986; Van Damme *et al.*, 1987; Van Damme and Peumans, 1988).

History of medicinal uses of *Urtica*

Egyptians

There are references of the medicinal use of *Urtica* plants dating back to over 2,000 years and possibly over 3,000 years ago. In ancient times there were reports of the use of a nettle infusion for the relief of arthritis and lumbago pains.

In *Healing Herbs of the Bible*, Harrison reports that during the Egyptian era (.1820 BCE–CE 250) *U. pilulifera* was included in the Passover herbs, and an infusion of this plant was given to treat the pains of arthritis and lumbago (Harrison, 1966). There are four species of nettle that still occur in Palestine: *U. dioica, U. pilulifera, U. urens, and U. caudate* (Harrison, 1966).

Greeks and Romans

The ancient Greeks had an extensive range of plants which they used for medicinal purposes. Nettle was a common remedy suggested for many different ailments. Theophrastus (*c.*300 BCE), was a pupil of Aristotle, and is often described as 'The Father of Botany'. His *De historia et causis plantarum* is the earliest work of scientific botany. Part of this book is devoted to plant-lore and the gathering of plants for medicinal purposes. He collated and systematized the existing botanical knowledge and described about 500 plants. Nettle was recommended as a potherb, and the juice was prescribed raw or cooked along with rhubarb and garden nightshade (containing atropine, hyoscine, and hyoscyamine), for those who needed its warming action (Hort, 1916).

The Roman Pliny the Elder (CE 23–79) listed the Greek plant medicinal remedies in *The Natural History of Pliny*. He recounted how Hippocrates of Cos, in the third century BCE, had described five varieties of nettle: *U. dioica, U. urens, U. pilulifera, Lamium* (similar in appearance but unrelated to *Urtica*) and a Herculaneum nettle. This latter nettle was said to give off a smell, but has not been identified in modern times. Pliny recounted the Greek recommendation of nettle as a nutritious food in the spring, which was said to prevent disease throughout the year. The custom of eating fresh nettles in the spring has continued in the twentieth century in Sweden and rural areas of Great Britain.

Hippocrates and his followers reported 61 remedies using the nettle. His follower Nicander prescribed nettle seed as an antidote to the poison of hemlock, fungi, and quicksilver (mercury). Apollodorus, another Hippocratic student, used it in the broth of boiled tortoise as a treatment for the bite of the salamander, and it was also used as an antidote for the poison of henbane, serpents, and scorpions. He also prescribed nettle decoction for a bruised uvula, prolapsed uterus, and anal prolapse. He suggested stinging of the legs and forehead for lethargy, the topical application of nettles (particularly root extract) for nose bleeds, and mixed with salt for dog bites, cancers, foul ulcers, sprains, and inflamed tumours (Bostock and Riley, 1890; Castleman, 1991).

The topical application of nettle seed was said to arrest mucous discharge from the nostrils. An electuary (paste) of nettle seed and honey was said to relieve breathing difficulties and clear the chest by expectoration. Hippocrates also recommended nettle seed in sweet wine to alleviate pain and as a 'purgative of the uterus', nettle seed with salt to expel intestinal worms, and nettle seed liniment to restore hair loss. He also suggested nettle seed applied topically with 'old oil', or leaves beaten up with bear's grease, for gout and other joint diseases. The nettle seed of *U. pilulifera* (Roman nettle) from Alexandria was said to be the most powerful (Adams, 1985).

The Greek physician and botanist Dioscorides, who travelled with the Roman armies, wrote his extensive *Materia Medica* in CE 77–78. The only surviving manuscript of Dioscorides' *Materia Medica* is in the *Juliana Anicia Codex* (CE 512) housed in the Austrian National Library in Vienna. This is known as the *Vienna Dioscorides*, and is often said to be the oldest and most valuable work in the history of botany and pharmacology. Dioscorides recommended both *U. pilulifera* and *U. urens* (small nettle) for various ailments. He advised that the leaves of either nettle should be applied with salt for dog bites, ulcers, malignant tumours, and for pain. The advice for the treatment of nosebleeds was the application of crushed leaves to the nostrils. Nettle leaves mixed with small shellfish were prescribed for stomach upsets and as a diuretic. Nettle juice was suggested as a gargle for an inflamed uvula (Dioscorides, CE 77; Gunther, 1934; Swann, 1980).

The Roman soldiers are said to have brought their own nettle, *U. pilulifera,* to the United Kingdom to treat their tired painful legs on long marches in the cold and wet British climate. They are said to have flagellated their legs to warm themselves and treat their rheumatism (Duke, 1983; Grieve, 1992). The Roman nettle (*U. pilulifera*) is probably now extinct in Britain but still grows in southern Europe (Hastings, 1997), and is claimed by some authors to have a fiercer sting than the common British nettle *U. dioica* (Bombardelli and Morazzoni, 1997).

In the second century CE, Galen, the Greek physician, who also practised in Rome, wrote *De Simplicibus*. He recommended nettle leaves for the following uses: as a diuretic and laxative, for dog bites, gangrenous wounds, swellings, nose bleeding, for 'relieving menstruation', for the treatment of 'spleen related illness', pleurisy, pneumonia, asthma, tinea, and mouth sores (Bombardelli and Morazzoni, 1997; Blumenthal, 2000).

Two hundred years after Galen, Apuleius Platonicus (*c.*CE 400), wrote his *Herbarium of Apuleius*. He suggested using the nettle combined with hemp or cannabis to treat symptoms of feeling cold after being burnt (shocked), and nettle by itself for 'cold injury'. Nettle was also listed as a remedy for swellings, bruised wounds, sore joints, foul and rotten wounds, and 'for a woman's flux' (Cockayne, 1865).

Dark Ages (fifth to tenth centuries)

In his translation of the earliest medical treatise of western European Anglo-Saxon medicine in the lacuna of the Dark Ages (fifth to tenth centuries), *Leechdoms, Wort-cunning and Starcraft of Early England*, Oswald Cockayne reports many ancient uses of nettles. 'Withered feet and legs' were treated with *U. urens* as follows:

> in the case of many a man, his feet shrink up to his hams, work baths, add tares and cress and small nettle and beewort (*Acorus calamus*) put hot stones well heated in a trough, warm the hams with the stone bath, when they are in sweat, then let him, the patient, duly arrange the bones as well as he can, apply a splint, and it is so much the better the oftener a man bathes with the preparation.
>
> (Oswald Cockayne, 1865)

Nettle was suggested for shingles (herpes zoster) as one of the many ingredients of a decoction of several herbs. The patient was then given this to drink for ten nights, followed by a drink of mistletoe (now known to be poisonous) in wine for nine days without meat to eat. Two forms of salve (soothing ointments) were made from nettle. A wound salve was made of a mixture of woad (*Isatis tinctoria* – yellow flowered plant of cabbage family which produces a blue dye when dried) and of nettle, pounded well, boiled in butter, strained through a cloth, white salt added and the mixture shaken thoroughly. A wen (sebaceous cyst) salve, and for wen boils, was made by a mixture of four herbs: nettle, hemlock, wenwort (probably *Ranunculus ficaria*), and the small moor-wort, all boiled in butter and sheep's grease. This was then mixed with ships tar, garlic, cropleek, sedgeleek (*Allium schoenoprasum*), and salt. The resulting 'salve' was then warmed on a cloth and applied to the wen (Cockayne, 1865).

Nettle was suggested for constipation when used in wine, in oil, or sodden in water with salt. There was also a remedy ('Woden's Charm') for 'dry disease', which probably involved problems with the sinuses or lungs, mucous membranes, and skin. The remedy was a mixture of nine herbs (one a tree) that was said to be sacred to Woden (God of wisdom) and consisted of nettle, mugwort, plantain, watercress, camomile, chervil, fennel, crab apple, and 'atterlothe' an unidentified plant, all mixed in ale. Neck pain was treated with the lower leaves of nettle boiled in ox fat and butter; and the 'thighs' smeared with the salve, while for thigh pain the 'neck' was smeared with nettle salve! Also for neck pain lower nettle leaves were boiled in vinegar, ox gall was added and then the wort (nettle) removed, and the decoction was smeared on the neck (Cockayne, 1865).

Twelfth century

In the twelfth century, Saint Hildegard from Bingen, Germany (1098–1179) is reported to have recommended the use of nettle seeds for stomach ache (Bombardelli and Morazzoni, 1997).

Seventeenth century

Five hundred years later, in the seventeenth century, T. Johnson wrote of the virtues of nettles, suggesting them as counter poison for henbane, serpents, scorpions, and the application of nettle juice for the sting of the nettle (Woodward, 1927). These were almost unchanged from similar remedies described by Galen in the second century CE.

Culpeper (1616–1654), the astrologer-physician, claimed nettle leaves or roots boiled and mixed with honey and sugar produced a safe and useful medicine for wheezing and lung congestion. He also recommended a nettle/honey extract as a gargle for throat and mouth infections, and claimed that nettles were helpful for bladder stones/gravel, worms in children, antiseptic wash for wounds and skin infections, and as an antidote to venomous stings from animals. His remedy for gout, sciatica, or joint aches was 'nettle leaves mixed with wallwort or deanwort, bruised and applied directly to the painful part' (Culpeper, 1653).

Also in the seventeenth century, Bock and Matthiolus recommended nettle leaves as a diuretic, an aphrodisiac, an anti-haemorrhagic, for wound healing, 'against swellings and cancer', and in the treatment of kidney disease. 'Urtication' or 'flogging' with nettles was prescribed for chronic rheumatism, lethargy, coma, paralysis, and also in the treatment of typhus and cholera (Bombaradelli and Morazzoni, 1997).

Eighteenth century

During the eighteenth century Elizabeth Blackwell in her beautifully illustrated *Curious Herbal* noted the value of nettle for cooking, as a restringent, and its use against all kinds of 'inward bleeding'. Nettle juice was recommended for local application to bleeding noses or other wounds, the root as a diuretic and treatment for jaundice, and nettle seed for coughs and shortness of breath. In his publication *Primitive Physic*, John Wesley, the preacher, recommended nettles as an anti-haemorrhagic using bruised nettle tops applied directly to bleeding wounds, nettle juice decoction for internal bleeding, and for violent bleeding piles two ounces of lightly boiled nettle juice with a little sugar. For hoarseness he advised dried roots powdered and mixed with treacle, and for jaundice a strong decoction taken twice daily. He also suggested eating nettles for scurvy, pleurisy and worms, and applying nettle juice directly to the stinging nettle rash. He advised boiled nettles for white swelling of joints, and as a cure for sciatica he suggested a fomentation of boiled nettles applied directly as a poultice. Wesley also proclaimed the value of nettle in old age, 'a decoction of nettles, will probably renew their strength for some years' (Wesley, 1792).

Nineteenth century

The nineteenth century physicians recommended nettles for many conditions. Thornton suggested nettle juice treatment for scurvy, haemoptysis, nosebleeds (local application), and in large doses for cancer. Nettle seed oil internally was said to be good for 'les plaisirs de l'amour' but the effect was 'very forcing and to be cautiously employed' (Thornton, 1810)! Wren lists actions of nettle as a diuretic, a nettle rash remedy, a tonic botanic beer, and locally as an astringent. Nettle seeds or herb in boiling water were prescribed for consumption (tuberculosis) (Wren, 1877). Phelps Brown suggests nettle internally as a diuretic, a tonic, a remedy for dysentery, haemorrhoids, bladder and kidney stones, and the seeds and flowers in wine for fevers. In addition, the local application of nettle leaves was recommended as a styptic astringent for bleeding wounds, and a rubefacient for arthritis and rheumatism (Phelps Brown, 1885). The nineteenth century Eclectics (physicians who chose the 'best' from various disciplines) recommended nettle primarily as a diuretic to treat urinary, bladder, and kidney problems (Fluk, 1976). They also used it for its styptic properties in the treatment of bleeding haemorrhoids, in addition to its use for infant diarrhoea and eczema (Castleman, 1991).

Twentieth century

There are numerous references in the twentieth century literature to the use of nettle for food and medicinal purposes. Its reputation for being a highly nutritious food with restorative powers for the sick has been particularly appreciated in poor rural areas, especially since it is freely available from the hedgerows and fields (Lanska, 1992; Vickery, 1993). In Scotland it has been traditionally used as a food, and as a strong root extract in nettle ale (beer) as 'a cure for jaundice' (Beith, 1995).

Nettle beer recipe

> Nettle Beer was popular in the United Kingdom 40–50 years ago as a refreshing drink and well-respected remedy for rheumatic aches and pains. Recipe: Into two gallons of water put two gallons dry measure of young nettle tops, four cupfuls of young dandelion leaves, four cupfuls of goose grass and one ounce of root ginger. Bring to the boil and simmer for half an hour. Then strain and add one and a half pounds of castor sugar, stir until the sugar is dissolved and add the juice of a lemon. Beat one ounce of yeast into a cream and add this to the still warm mixture. Let it ferment for 4–5 hours in a warm place then bottle it, using corks. It is ready to drink straight away but becomes more alcoholic if left to ferment for any length of time.
>
> (Autolycus, 1978)

Traditional herbal remedies

Nettle is a traditional remedy for anaemia and 'lack of energy' (Grieve, 1992), its claimed restorative powers are probably related to its high level of iron, vitamin C, and other nutrients. Roy Vickery in *A Dictionary of Plant-Lore* writes informatively giving many references to the stinging nettle (Vickery, 1995).

He lists references of medicinal claims including direct application of stinging nettle or nettle ointment to the affected rheumatic part. In the Western Isles of Scotland nettle leaves were taken internally for rheumatics, indigestion, coughs, and as a general appetizer. Only young shoots were considered suitable for eating (Beith, 1995).

In *A Modern Herbal*, Mary Grieve lists the following traditional medicinal uses of nettle: an arrester of bleeding for nosebleeds (lint with nettle juice), as a drink for internal haemorrhage of the stomach/lungs/or nose and nettle rash, and a soothing application to burns or cuts. Also it is claimed to be a homeopathic remedy as a tincture used internally for rheumatic gout, chicken pox, nettle rash, and externally for bruises. She also recommends 'urtication or flogging' with nettles for chronic rheumatism and loss of muscle power. Herbal dried nettles (usually taken as tea or capsules) are suggested for asthma, goitre, ague (fever), and kidney conditions 'but it is a doubtful diuretic'. Nettle is said to be a common traditional ingredient of hair lotions, tonics, and shampoos (Grieve, 1992).

Modern medicinal uses of *Urtica*

Introduction

In this section present day usage of nettle by herbalists and others, based on historical and anecdotal evidence, will be considered. The prescribing of nettle based on scientific evidence will be presented in Chapter 4, 'Various therapeutic uses of *Urtica*'.

Local application of nettles

The sting of the nettle leaf (urtication) is used in the treatment of joint pains and neuralgia. There have been published reports of this mode of use in Australia (Czarnetski *et al.*, 1990), Germany (Weiss, 1988), India (Shah and Joshi, 1971), North America (Duke, 1997), South America (Brisley, 1994), Northern Italy (Cappelletti *et al.*, 1982), and the United Kingdom (Thomson, 1976; Mills, 1991; Bradley, 1992; Randall, 1994; Vickery, 1995; Mabey, 1996; Randall *et al.*, 1999, 2000). The author has also received personal communications of the use of urtication with *U. dioica* for arthritis in France, New Zealand, and Poland.

The Thompson (Native) Indians of Vancouver Island in Canada still use this traditional therapy for rheumatoid arthritis, and other muscloskeletal pains (Turner *et al.*, 1990). Ecuador native Indians are also reported to use nettle urtication (stinging) to treat muscle and joint pains, and muscular aches on long marches (Brisley, 1994). In two ethno-botanical studies in India *Urtica* has been reported to be used for sprains and swellings (*U. dioica* and *U. parviflora* Roxb.) by direct application of branches of leaves, and for gout and rheumatic pains by external application of boiled *Urtica dioica* leaves (Shah and Joshi, 1971; Lal and Yadav, 1983).

Several methods of local treatment of painful joints with *Urtica* have been reported: by beating with a bunch of nettles (Pahlow, 1992; Vickery, 1995), direct application of the stinging leaf (Schauenberg and Paris, 1977; Lust, 1983; Mills, 1991; Stary, 1991; Bradley, 1992; Duke, 1997; Randall *et al.*, 1999, 2000; Randall, 2001), nettle poultice of green steeped leaves (Kloss, 1985), nettle juice (Cyran, 1981), nettle ointment of chopped nettles mixed with salt, vinegar, and lard (Vickery, 1995), nettle spirit/tincture (Weiss, 1988); and boiled leaves (Lal and Yadav, 1983).

Dr Rudolf F. Weiss in Germany recommends whipping with a nettle switch or the external application of nettle spirit. He suggests whipping the painful area with a bunch (switch) of nettles for three successive days, then having a break of three days to prevent sensitization. Patients are advised to avoid cold water touching the treated skin for the rest of the day, as this is said to cause the normal pleasurable warmth to turn into an unpleasant burning sensation (Pahlow, 1992). Weiss prescribes the use of nettle spirit externally which is said to 'utilise the local hyperaemic action and the irritant effect achieved via the cutaneous nerves'. The nettle spirit is applied to the painful area for neuralgic and rheumatic pain, particularly with degenerative and chronic arthritic conditions such as lumbago, sciatica pain, chronic tendonitis, sprains, and other similar conditions (Weiss, 1988).

An ointment made from *U. urens* has also been reported as effective for back pain (personal communication, from Surrey, United Kingdom). Anecdotal reports from Canada of the use of nettle lotion for application to painful joints have been communicated to the author. Nettle juice and dried powdered nettle leaf have been used as an astringent for nose bleeds. A lotion of stewed nettle liquor has been used as a soothing antiseptic wash for infected wounds (Schauenberg and Paris, 1977), burns, cuts, and eczema (Wren, 1988). Russian studies are said to show that nettle tea has a local antibacterial activity, and mouthwashes – especially when combined with Juniper – have been said to reduce dental plaque and gingivitis. There are however no scientific trials to support these claims, and one study on a mouth rinse of three herbal extracts (*U. dioica, Juniperis communis*, and *Achillae millefolium*) demonstrated no effect on dental plaque growth and gingival health (VanderWeijden *et al.*, 1998). Interestingly the North American Native Indians traditionally use nettle extracts applied directly to the gums for dental pain (Duke, 1997).

The entire nettle plant boiled in vinegar and water and preserved in cologne produces the world-famed Kneipp hair remedy (Lust, 1983).

Nettle taken as an internal medication

Nettle leaves

Present day medicinal use of nettle leaves taken internally is mainly restricted to dried leaves in nettle tea or capsules of freeze-dried nettle leaf, or as a decoction (from stewed nettles).

Nettle tea is recommended for 'breast, lung, stomach, and urinary problems' by Lust (1983). Nettle is a constituent of 'anti-diabetic' teas, which are recommended by some herbalists. However, nettle is variously reported as both increasing (Oliver-Bever and

Zahland, 1979) and decreasing blood sugar levels (Wren, 1988; Newall *et al.*, 1996), and the use of these teas by diabetic patients is not recommended (Bissett, 1994). The traditional remedy of nettle tea for gout and rheumatism is still suggested by herbalists, and there is claimed to be some experimental evidence of it causing increased excretion of uric acid in the urine (Mills, 1988). In Germany nettle tea is used as a mild diuretic; 3–4 teaspoonfuls of the dried leaf in 150 ml of boiling water is taken three times a day, but it is not considered powerful enough for hypertension or cardiac oedema (Tyler, 1994). *Urtica* is called 'Shisoon' and 'Bichhubooti' in north-eastern India and said to have magical powers. The plant is used to guard against evil spirits; when newly wed brides or sick infants move from one place to another, they are always moved with a branch of *Urtica*. These plants were traditionally used as decoration of the walls during the magical-religious ceremony called a Jagar, when the anger of the household god is assuaged in an attempt to cure the ill member of the family (Shah and Joshi, 1971). This process is observed to cure 50% of patients, rather better than the usual 30% expected from a placebo response. Nettle tea made from fresh nettle tops is known as Bichu, Bichhubooti, Chicru, or Shisoon in India, and is prescribed for intermittent fever, 'gravel in the kidneys' and 'excessive menstrual flow' (Dasiur, 1962; Shah and Joshi, 1971).

In an ethno-botanical study by Baranov of plants used in Manchuria for wild vegetable use, *U. augustifoloia*, *U. lactevirens*, and *U. cannabina* are listed (Baranov, 1967). *Urtica cannabina* is said to have a very powerful sting (Hastings, 1997). The Chinese ethnic populations of Manchuria are said to have received their essential knowledge of wild vegetable plants from their fathers who brought it from China proper (Baranov, 1967). Duke and Ayensu in their *Medicinal Plants of China* list six Urticaceae but include none of the *Urtica* genus. In particular the stinging Nilgiri nettle (*Girardinia palmate*) is listed drunk with wine for malignant boils and cooked with pork for stomach ache (Duke and Ayensu, 1985). *Urtica thunbergiana* has been listed for its use in China as a haemostatic for menorrhagia, as a galactalogue (increases breast milk production), as a treatment for snakebite and pemphigus (rare condition of blisters of skin, mouth, and mucous membranes), and as an anti-emetic (Read and Liu, 1927). In a more recent publication on Chinese plant remedies Zhang lists the following species of *Urtica* for their medicinal use: *U. angustifolia*, *U. cannabina*, *U. dioica*, *U. fissa*, *U. hyperborean*, *U. laetevirens*, *U. macrorrhiza*, *U. mairei*, *U. thunbergiana*, *U. tibetica*, *and U. triangularis* (Zhang, 1994).

Boiled leaves are still used mainly in rural areas throughout Europe as a highly nutrient vegetable rich in vitamin C, iron, magnesium, and other minerals (Castleman, 1991; Vickery, 1995; Duke, 1997). Its high levels of iron and vitamin C confirm the rationale for its traditionally proclaimed value in the treatment of anaemia and scurvy (Sapronova, 1989; Stary, 1991; Patten, 1993). As described earlier in this chapter it is claimed that nettle leaves increase milk production of cows (Grieve, 1992) and humans (Mills, 1988).

In western Europe nettle decoctions or infusions are used as a diuretic, anti-haemorrhagic, anti-rheumatic, for allergic skin rashes and asthma, and for gastric and biliary conditions (Dorfler and Roselt, 1986; Grieve, 1992; Barnes and Ernst, 1998).

In Russia nettles are also traditionally used for gall bladder and liver ailments (Geltman, *c*.1995).

Nettle root

Herbalists use nettle root occasionally in the United Kingdom for 'prostatic' conditions (Bradley, 1996), but it is used much more extensively in Germany where phytotherapy is more widely recognized and used.

Conclusions

Nettles have been used for over two thousand years in various ways including the production of fibre for fabrics, threads, and paper. They are also used as animal fodder, a dye for fabrics, and as a vegetarian rennet in cheese making.

The main medicinal uses of nettles historically were internally as a tonic and highly nutrient food, and treatment of the following conditions – anaemia, rheumatism and arthritis, eczema and asthma, urinary 'gravel' and stomach complaints, skin infections, and as an 'anti-haemorrhagic'.

Nettles have also been traditionally used externally as a hair tonic/shampoo, locally applied styptic for nosebleeds and haemorrhoids, and stinging treatment (urtication) for arthritis and rheumatism.

References

Adams, F. (1985) *The Genuine Works of Hippocrates*. (Original edition 1849.) The Sydenham Society, London.

Ahket, G. (1983) *Herbal Treatment of Common Ailments*, Lloyd O'Neill Pty Limited, Melbourne.

Autolycus (1978) *The Field*, 5th April, UK.

Baranov, A. I. (1967) Wild vegetables of the Chinese in Manchuria. *Economic Botany*, 21, 140–55.

Barnes, J. and Ernst, E. (1998) Traditional herbalists' prescriptions for common clinical conditions: a survey of members of the UK National Institute of Medical Herbalists. *Phytotherapy Research*, 12, 369–71.

Beith, M. (1995) *Healing Threads – Traditional Medicines of the Scottish Highlands and Islands*. Polygon Press, p. 214.

Bisset, N. G. (1994) *Herbal Drugs and Phytopharmaceuticals*. Medpharm, Stuttgart, pp. 502–9.

Blackwell, E. (1737) *A Curious Herbal*, Vol. 1. John Nourse, London.

Blumenthal, M. (2000) *Herbal Medicine: Expanded Commission E Monograph*. American Botanical Council, Austin, p. 387.

Bombardelli, E. and Morazzoni, P. (1997) *Urtica dioica* L. *Fitoterapia*, 67(5), 387–402.

Bostock, J. and Riley, H. T. (1890) *The Natural History of Pliny* (translation) Vol. 4. George Bell, London.

Bradley, P. (1992) *British Herbal Compendium* (1). British Herbal Medicine Association, Exeter England, pp. 166–7.

Bradley, P. (1996) *British Herbal Pharmacopoeia*, Fourth edition. British Herbal Medicine Association, Exeter England, pp. 142–4.

Brisley, G. (1994) Ecuador Indians' use of nettle sting for joint pains (letter). *GP*, Dec. 9, 289.

Cappelletti, E. M., Trevisan, R., and Caniato, R. (1982) External antirheumatics from herbs in Northern Italy. *Journal of Ethnopharmacology*, 6(2), 161–90.

Castleman, M. (1991) *The Healing Herbs*. Schartz Books, Melbourne.

Cockayne, O. (1865) *Leech Book of the Bald* (*c*.AD 925) translated by Cockayne O. *Leechdoms, Wort-cunning and Starcraft of Early England*. London, Vol. 2.

Culpeper, N. (1653) *Culpeper's British Herbal* 1934. Milner & Co. Ltd. London.

Cyran, H. B. (1981) *The Australian and New Zealand Book of Herbs*. Child and Henry Publishing, Sydney.

Czarnetzki, B. M., Thiele, T., and Rosenbach, T. (1990) Immunoreactive leukotrienes in nettle plants (*Urtica urens*). *Int Arch Allergy Immunol*, 91, 43–6.

Dasiur, J. F. (1962) *Medicinal Plants of India and Pakistan*. Tarapoteuala, Bombay.

Dioscorides (CE 77) *Materia Medica Book 4*, chapter 93 – Nettles and their uses. Only copy, Juliana Anicia codex, Vienna codex (AD 512). Austrian National Library, Vienna.

Dorfler, P. H. and Roselt, G. (1986) *Dictionary of Healing Plants*. Blandford Press, London.

Duke, J. A. (1983) *Medicinal Plants of the Bible*. Trado-Medic Books, New York, p. 161.

Duke, J. A. (1997) *The Green Pharmacy*. CRC Press, Florida, p. 53.

Duke, J. A. and Ayensu E. S. (1985) *Medicinal Plants of China*. Reference Publications, Algonac Michigan.

Fluk, H. (1976) *Medicinal Plants*. W. Foulsham, England.

Geltman, D. (*c.* 1995) *Urtica* – PhD thesis. Institute of Science, St Petersburg.

Grieve, M. (1983) *A Modern Herbal*. Penguin Books, London.

Grieve, M. (1992) *A Modern Herbal*. Hafner Press, New York, p. 574.

Gunther, R. T. (1934) *The Greek Herbal of Dioscorides* translated by Goodyer J. Book IV, No. 94. Oxford University Press, Oxford, p. 491.

Harrison, R. K. (1966) *Healing Herbs of the Bible*. E.J. Brill, Leiden, p. 29.

Hastings, L. (1997) Economic Botanist, Royal Botanic Gardens Kew, personal communication.

Hort, Sir A. F. (1916) *Enquiry into Plants and Minor Works on Odours and Weather Signs*, Translation of Theophrastus (*c.* 300 BC). Book VII. Heinemann, London, p. 105.

Kloss, J. (1985) *Back to Eden*. Back to Eden Books, Loma Linda California.

Lal, D. S. and Yadav, B. K. (1983) Folk medicines of Kurukshetra district (Haryana), India. *Economic Botany*, 37, 299–305.

Lanska, D. (1992) *Illustrated Guide to Edible Plants*. Chacellor Press, London, p. 206.

Lust, B. (1983) *About Herbs, Medicines from the Meadows*. Thorsons Pub., London.

Mabey, R. (1996) *Flora Britannica*. Sinclair Stevenson, London, p. 68.

Mills, S. Y. (1988) *The Dictionary of Modern Herbalism*. Healing Art Press, Rochester Vermont.

Mills, S. Y. (1991) *The Essential Book of Herbal Medicine*. Penguin, London.

Newall, C. A., Anderson, L. A., and Phillipson, J. D. (1996) *Herbal Medicines: A Guide for Health-care Professionals*. The Pharmaceutical Press, London.

Oliver-Bever, B. and Zahland, G. R. (1979) Plants with oral hypoglycemic activity. *Quart J Crude Drug Res*, 17, 139–96.

Pahlow, M. (1992) *Healing Plants* (translation from German). Baron's Educational series, Hauppauge, New York, p. 72.

Patten, G. (1993) Urtica. *Austr J Med Herbalism*, 5(1), 5–13.

Peumanns, W. J., De Ley, M., and Broekaert, W. F. (1984) *FEBS Lett*, 177, 99.

Phelps Brown, O. (1885) *The Complete Herbalist*. Phelps Brown, O. London, 126–7.

Randall, C. F. (1994) Stinging nettles for osteoarthritis pain of the hip (letter). *Br J Gen Pract*, 44, 533–4.

Randall, C. F. (2001) *Treatment of Musculoskeletal Pain with the Sting of the Stinging Nettle: Urtica dioica.* MD thesis. University of Plymouth.

Randall, C. F. (2002) *Personal Observation by Author, during Visit to Cheese Factory.*

Randall, C. F., Meethan, K., Randall, H., and Dobbs, F. (1999) Nettle sting of *Urtica dioica* for joint pain – an exploratory study of this complementary therapy. *Complementary Therapies in Medicine*, 7, 126–31.

Randall, C. F., Randall, H., Hutton, C., Sanders, H., and Dobbs, F. (2000) Randomized controlled trial of nettle sting for treatment of base-of-thumb pain. *J Royal Soc Med*, 93, 305–9.

Read, B. E. and Liu, J. C. (1927) *Flora Sinensis Series A, Vol.1 Plantae Medicinalis Sinensis*, second edition. Department of pharmacology, Peking union medical college, in collaboration with Peking laboratory of Natural History, Peking China.

Robertson, G. (1998) *Personal Communication Concerning Use of Nettles as Food in Sweden*, University of Plymouth.

Sapronova, N. N. (1989) The content of vitamin K and certain microelements in *Urtica dioica*. *Rastitel'nye – Resrsy*, 25, 243–7.

Schauenberg, P. and Paris, F. (1977) *Guide to Medicinal Plants*. Lutterworth Press, London.

Shah, N. C. and Joshi, M. C. (1971) An ethnobotanical study of the Kumaon region of India. *Economic Botany*, 25, 414–22.

Shibuya, N., Goldstein, I. J., Shafer, J. A., Peumanns, W. J., and Broekaert, W. F. (1986) *Arch Biochem Biophys*, 249, 215.

Stary, F. (1991) *Medicinal Herbs and Plants*. Treasure Press, London.

Swann, C. (1980) *Nettles Healers of the Wild*. Thorson's Publishers Ltd., Wellingborough Northants, UK, p. 9.

Thomson, W. (1976) *Herbs That Heal*. Adam and Charles Black, London.

Thornton, R. J. (1810) *A New Family Herbal*. Richard Phillips, London.

Turner, N. J., Thompson, L. C., and Thompson, M. T. (1990) *Thompson Ethnobotany – Memoir 3*. Published by Royal British Columbia Museum, Victoria, B.C. Canada, p. 289.

Tyler, V. E. (1994) *Herbs of Choice: The Therapeutic use of Phytomedicinals*. Pharmaceutical Products Press, New York, pp. 84–5.

Van Damme, E. J., Broekaert, W. F., and Peumanns, W. J. (1987) *Plant Physiol*, 86, 598.

Van Damme, E. J. and Peumanns, W. J. (1988) *Plant Physiol*, 71, 328.

VanderWeijden, G. A., Timmer, C. J., Timmerman, M. F., Reijerse, E., Mantel, M. S., and VanderWelden, U. (1998) The effect of herbal extracts in an experimental mouth rinse on established plaque and gingivitis. *J Clin Peroidontology*, 25(5), 413–6.

Vickery, R. (1993) Nettles; their uses and folklore in the British Isles. *Folk Life*, 31, 88–93.

Vickery, R. (1995) *A Dictionary of Plant-Lore*. Oxford University Press, Oxford, pp. 253–8.

Weiss, R. F. (1991) *Herbal Medicine* (English Translation-Meuss AR). Beaconsfield Pub. Ltd, Beaconsfield England.

Wesley, J. (1792) *Primitive Physic*. 24th edn, G. Paramore, London.

Woodward, M. (1927) *Gerard's Herbal* by Th. Johnson, edited version of first edition (1636).

Wren, R. C. (1877) *Potter's Cyclopaedia of Botanical Drugs and Preparations*. 2nd edn, Potter & Clarke Ltd, London.

Wren, C. (1988) *Potters New Cyclopaedia of Botanical Drugs and Preparations*. Health Sciences Press, Bradford England.

Zhang, H. K. (1994) *Synopsis of Traditional Chinese Medicine Sources*, edited by Zhang Hui Yuan and Zhang Zhi Ying. Science Press, Beijing.

3   The chemical and pharmacological aspects of *Urtica*

Gulsel Kavalali

Introduction

In a crude drug, it is important to know the structure of the chemical compounds which are responsible for its biological activities. It is also important to know the other inactive compounds present in the active fraction that may improve solubility, bioavailability and act as a synergist of the major active compounds.

The recognition of the medical importance of *Urtica* species started in the early twentieth century. Since then, considerable progress has been made in finding out the structure of the active compounds owing to the improvements in separation techniques and spectroscopic methods of structure elucidation. The constituents of *Urtica dioica* are of medical interest, as extracts of the roots and leaves are widely used in traditional medicine in many areas of the world. The major traditional uses of *Urtica* are based on its anti-inflammatory, anti-hyperplastic (prostate), insecticide, antifungal and selective antiviral activities. *Urtica* species are also used in many phytotherapeutic preparations.

This chapter discusses the scientific evidence of *in vitro* and animal studies on the biological activity of *Urtica* species and its constituents.

Chemical studies

Constituents of Urtica dioica

Nettle irritation arises from the fluid produced at the base of the stinging hair. Flury (1927) reports that the urticating hair contains a small amount of formic acid. However, although formic acid may play a part in causing the burning sensation, it is a minor factor. The chemical composition of urticating fluid of the nettle was investigated by Emmelin and Feldberg (1947). These workers found the fluid in the hairs of *U. dioica* contains at least three compounds which can evoke smooth muscle reactions. One of these is acetylcholine and the other one is probably histamine. The third substance has not been identified yet but the authors described six of its properties. Collier and Chesher (1956) reported another active compound, 5-Hydroxy Tryptamine (5-HT), which was shown to stimulate the rat uterus *in vitro*.

The active chemical portion of *U. dioica* comprises of approximately 50 different structurally elucidated compounds from the lipophilic fraction. These include sterines, triterpenic acids, coumarins, simple phenols, lignans, ceramides and hydroxy fatty acids (Chaurasia and Wichtl, 1986a,b, 1987a,b; Kraus and Spiteller, 1991).

Chaurasia and Wichtl (1987a) isolated flavonol glycosides for the first time from the flowers of *U. dioica* (male and female) and identified their structure by chromatographic and spectroscopic methods (UV, FD-MS, ^{1}H-NMR, ^{13}C-NMR). *Urtica dioica* leaf was shown to contain chlorophyll α and β, coproporphyrin, protoporphyrin, β-carotene, violaxanthin, xantophyll, zeaxanthin luteoxanthin, ascorbic acid, pantothenic acid, choline, histamine, 5-hydroxy tryptamine, nicotine, threono-1,4-lactone and formic, acetic, butyric, citric, *p*-coumaric, succinic and threonic acids (Chaurasia, 1987b; Schomaker *et al.*, 1995) and tannin (Kraus and Spiteller, 1991; Anikina, 1996).

A neutral and an acidic carbohydrate–protein polymer was isolated from the leaves of *U. dioica* (Anderson and Wold, 1978). This neutral fraction was a glycoprotein containing the serine-O-galactoside glycopeptide bond.

Galacturonic acid was the major component of the acidic fraction. A trigalactosyl diglyceride with antigenic activity was also isolated (Radunz, 1976) as well as kaempferol, isorhamnatin, quercetin and their glycosides (Wojtaszek *et al.*, 1986; Karakaya and Nehir El, 1999). Lipase was found to be present in the leaves and stems of *U. dioica* (Korchagina *et al.*, 1973). *Urtica* leaves have a relatively high level of protein (Hughes *et al.*, 1980), which is better quality in comparison with the proteins of many other green leafy vegetables (Ullrich and jahn-Dees bach, 1984).

The leaves of *Urtica* species are good sources of some important minerals and vitamins (Booth and Bradford, 1963; Adamski and Bieganska, 1980), especially of α-tocopherol and vitamin C (Hughes *et al.*, 1980). Silicon levels were determined in the leaves, stems and roots of *U. dioica* by Piekos and Paslawska (1976) who found silicon dioxide levels of the plant to range between 1.13% and 4.75% based on dry matter. The leaves were also noted for their particularly high content of some metals (Se, Zn and Fe, Mg) (Gravel *et al.*, 1994; Rzemykowska and Ostrowska, 1994; Salamon *et al.*, 2001). *Urtica dioica* (the common nettle) is a dioecious plant. Accumulation of flavonoids was comparable in both male and female forms. The content of polyphenolic acids both in the leaves and rhizomes was higher in the male form. The chemical composition of female polyphenolic acids was more complex (Weglarz and Roslon, 2000; Roslon and Weglarz, 2001).

Kavalali and Akcasu (1983) isolated an anti-coagulant from the leaves of *U. dioica*. This substance was confirmed as an heparinoid by IR, UV and NMR spectra. Kraus and Spiteller (1990, 1991a,b) detected terpene diols and terpene diol glucosides in the methanolic extract of *U. dioica*. Five new monoterpenoid components were elucidated by spectrometry as well as 18 phenolic compounds and eight lignans, some of which were previously unknown (Kraus and Spiteller, 1990).

Budzianowski (1991) found caffeoylmalic acid only in *U. dioica*. Trans-5-caffeoylquinic acid (chlorogenic acid) was found in both *U. dioica* and *U. urens*. *Urtica urens* also contains small amounts of 3- and 4-caffeoyl quinic acid. Smallman and Maneckjee (1981) found an acetylcholine synthesizing enzyme, choline acetyl-transferase in *U. dioica*, which appears to be the only plant to have this enzyme. It also contains a high level of acetylcholine in fresh hair and leaf tissue (15 and 88 μmol of acetylcholine/g fresh weight, respectively) (Emmelin and Feldberg, 1947). Barlow and Dixon (1973) found that the synthesis of acetylcholine in *U. dioica* continues throughout the life of the leaf.

The roots of *U. dioica* have been studied extensively and found to contain oleanol acid, 3-β-sitosterin and its derivatives and glycosides, scopoletin, homovanillyl alcohol, secoisolariciresinol glycoside, neo-olivil and its derivatives and glycosides (Schilcher and Effenberger, 1986; Chaurasia and Wichtl, 1986b; Chaurasia 1987).

Peumans *et al.* (1984) isolated a lectin: *U. dioica* agglutinin (UDA) from the roots of *U. dioica*. Lignans and their metabolites were isolated from the roots of *U. dioica* (Schottner *et al.*, 1997). Their affinity to human sex hormone binding globulin (SHBG) was tested in an *in vitro* assay in which all lignans except pinoresinol were found to develop a binding affinity to SHBG.

Wagner *et al.* (1989) isolated an immunologically active polysaccharide fraction which yielded 35% neutral sugar (glucose, galactose, rhamnose, mannose and xylose), 1% protein and 35% uronic acid.

Fractions of the methanolic extract of the roots of *U. dioica* were investigated by Gansser and Spiteller (1995a,b) for their *in vitro* inhibitory effect on aromatase, a key enzyme in steroid hormone metabolism. Four known compounds, secoisolariciresinol, oleanolic acid, ursolic acid and (9Z, 11E)-13 hydroxy-9,11-octadecadienoic acid and a new compound, 14-octacosanol, were isolated from these fractions and their structures were elucidated by chemical and spectral analysis.

Cytokinin type compounds, namely zeatin, zeatin nucleotide, dihydrozeatin, isopentanyl adenine, isopentanyl adenosine and isopentyl adenine nucleotide were detected in the xylem sap (Fusseder *et al.*, 1988). Many active constituents such as phytosterols, pentacyclin triterpenes, lignans coumarin, ceramides, hydroxy fatty acids were isolated from the lipophilic fraction. Polysaccharides and UDA were isolated from the hydrophilic fraction of the *Urtica* extract (Peumans *et al.*, 1984; Wagner and Willer, 1992) and are considered to be very important pharmacological findings.

Constituents of other species of Urtica dioica

There are only a few reports on the chemical composition of the other varieties of the *Urtica* species. However, the amine fraction (histamine, acetylcholine, choline and serotonin) is believed to be present in most of the varieties (Chaurasia, 1987; Schomakers *et al.*, 1995). Maitai *et al.* (1980) studied *U. massaica* species which grows in Kenya and found their chemical composition to be similar to that of *U. dioica*.

Histamine, acetylcholine and 5-hydroxy tryptamine were detected in *Urtica* species by several workers (Emmelin and Feldberg, 1947; Collier and Chesher, 1956; Mac Farlane, 1963; Saxena *et al.*, 1965). Saxena *et al.* (1965) also found these compounds to be present in *U. parviflore*. Hegnauer (1973) reported *U. dioica, U. ferox* Forst., *U. parviflora* Roxb., *U. urens* L., *U. pilulifera* L., *U. calophylla* Schlecht and *U. argentea* Forst. to contain acetylcholine, histamine and 5-hydroxy tryptamine. *U. argentina and U. cannabina* were both found to contain luteolin.

The smooth muscle stimulants present in *U. massaica* hairs were investigated by Maitai *et al.* (1980). The response to these stimulants in *U. massaica* hairs was partly blocked by atropine, suggesting the presence of acetylcholine and histamine in the hairs. These smooth muscle stimulating substances were surmised not to be present in dry hairs as dry hairs did not bring about the stinging sensation.

The hair and whole plant extracts of *U. urens* have been reported to contain high levels of leukotrienes (LTB$_4$ and LTC$_4$), which suggests that the mechanism of urtication after contact with *Urtica* is similar to that of insect venoms and cutaneous mast cell reactions with regard to its spectrum of mediators (Czarnetzki *et al.*, 1990).

Stachydrine was detected in *U. cannabina* (Vishnevskii and Proshunina, 1989) and the alkaloid bufotenin in *U. pilulifera* (Regula, 1972). Kavalali *et al.* (1993) studied *U. pilulifera* seed oil and found 70–76% linoleic acid and 12–25% oleic acid. Daniel and Wichtl (1991) studied rhizomes from *U. dioica* and *U. kioviensis* and found both species to contain β-sitosterin and sitosterin glycoside. Coumarin and scopoletin were found in *U. dioica*. *U. kioviensis* was found to contain a previously unknown compound with intense fluorescence at UV 365.

Plant lectins

Plant lectins (phytohaemagglutinins) are glycoproteins that bind sugar groups in a highly specific and selective way. Plant lectins are usually isolated from legumen seeds, although they may also be found in the roots, leaves and tubers (Majerus and Brodie, 1972). In 1988, Stillmark isolated phytohaemagglutinins for the first time from hundreds of different plants and classified them according to their molecular structure, biochemical properties and carbohydrate binding specificity.

All lectins have a very similar but not identical amino acid composition typified by high contents of aspartic acid, threonine, glycine, serine and leucine (Van Damme *et al.*, 1991). None of the lectins studied to date contain any amino sugar. They have only low levels of neutral sugars (mannose and glucose) and are able to bind carbohydrates. Their agglutination activity is usually inhibited by simple monosaccharides (mannose, glucose and their derivatives).

Lectins are detected and identified for their ability to agglutinate erythrocytes using modern protein chemistry techniques. Plant lectins were first obtained by using precipitation methods. Recently, ion exchange, gel filtration and affinity chromatographic methods are used. Plant lectins can play a role in plant defence. Cytotoxic lectins such as ricin, abrin, mistletoe, modeccin are strong poisons for all kinds of organism and can protect the plant against parasites and herbivores. The special chitin-binding lectin from *U. dioica* has antifungal properties (Peumans and Van Damme, 1993).

Urtica dioica *lectin (*Urtica dioica *Agglutinin, UDA)*

UDA is the first lectin isolated from *U. dioica*, a member of the Urticaceae family. It differs from all other plant lectins with respect to its molecular structure (Peumans *et al.*, 1984). It is present primarily in the rhizomes, roots and seeds but not in the leaves and stems of the plant (Lerner and Raikhel, 1992).

UDA is a small stable protein (MW 8.5 kDa) composed of 77 amino acids with high proportions of glycine, cysteine and tryptophan (Beintema and Peumans, 1992). It is specific for *N*-acetyl-glucosamine and has no covalently bound carbohydrate. It is heat and acid resistant, has a very low specific agglutination activity and also is the first known single chain lectin with no carbohydrate residue. UDA is a mixture of isolectins and has two carbohydrate binding sites per molecule consisting of a single polypeptide chain (Shibuya *et al.*, 1986). After 15 minutes boiling in water, 50% of the initial activity still remains. It is resistant to trypsin, 0.1N HCI, 1N CH$_3$COOH and 5% tri-chloro acetic acid (TCA). UDA causes non-specific agglutination of erythrocytes, induces γ-interferon in fresh human lymphocytes and has carbohydrate binding specificity (Van Damme *et al.*, 1988) which is three times more effective than wheat germ lectin (WGA). It was shown to possess both antifungal and insecticidal activity and to act synergistically with chitinase in inhibiting fungal growth (Broekaert *et al.*, 1989; Huesing *et al.*, 1991).

UDA has an immune modulatory effect on T lymphocytes in a dose dependent manner. It was shown to directly inhibit cell proliferation and block the binding of epidermal growth factor to its receptor on a tumour cell line (Wagner *et al.*, 1994). It induces production of γ-interferon in human lymphocytes (Peumans *et al.*, 1984). UDA is also a potent and selective inhibitor of the human immune deficiency virus (HIV-1 and HIV-2) and shows anti-prostatic activity by interfering with SHBG.

Pharmacological studies

In vitro *and animal studies*

Medicinal properties of *Urtica* have been reported by many authors. The primary use of *Urtica* species and their extracts in phytotherapy are concerned with their anti-inflammatory and benign prostatic hyperplasia activities which are discussed comprehensively in Chapter 4. This chapter deals with the scientific data from *in vitro* and animal studies on the biological activities of *Urtica* species and their constituents.

Anti-inflammatory activity

Peumans *et al.* (1984) first isolated a lectin from *U. dioica* rhizomes. It is the first single chain lectin to be found in plants. UDA differs from all other known plant lectins with respect to its molecular structure. It interacts with cells in a specific way and induces γ-interferon production by human lymphocytes.

Shibuya *et al.* (1986) found UDA has two carbohydrate binding sites per molecule chain. Van Damme *et al.* (1988) and Van Damme and Peumans (1987) isolated six isoform isolectins from the *Urtica* rhizomes. They all induce γ-interferon in fresh human lymphocytes.

Wagner *et al.* (1989) examined a polysaccharide fraction of the water extract of the roots of *U. dioica* and found anti-inflammatory activity in the carrageenan rat paw edema model and lymphocyte transformation test. They also found UDA to stimulate the proliferation of human lymphocytes.

Wagner and Willer (1990) isolated a lectin mixture and five chemically defined neutral and acidic polysaccharides from the *Urtica* roots. These compounds show immunological activities, T-lymphocyte stimulation and macrophage induced liberation of tumour necrosis factor (TNFα). These findings can offer an explanation for the antiphlogistic and antineoplastic activity of *Urtica* extracts.

Le Moal *et al.* (1992) reported in two papers that the lectin isolated by Peumans and co-workers (1984) from the *U. dioica* rhizomes was able to stimulate the proliferation of mouse thymocytes and spleen T lymphocytes. They compared UDA and Con A (Concanavalin A) and found that UDA had two- to three-fold lower activity than that induced by Con A. The kinetics of activation of UDA were different from T cell mitogen (Con A). Mikael *et al.* (1988) found the kinetics of this proliferation was markedly different from those of the classical T cell mitogen (Con A).

Galelli and Truffa-Bachi (1993) reported that UDA is different from the classical T cell lectin mitogens and is a superantigen without a pathogenicity factor. This was also shown in another study (Galelli *et al.*, 1995). Delcourt *et al.* (1996) found that UDA exhibits mitogenic features on mouse T cell and acts very differently from all the other superantigens on thymic lymphocytes. Rovira *et al.* (1999) reported major histocompatibility class I (MHC) molecules are present in UDA.

Obertreis *et al.* (1996) studied the antiphlogistic effects of an extract of *U. dioica* leaves. They isolated caffeoylmalic acid and tested its *in vitro* inhibitory effect on the biosynthesis of arachidonic acid metabolites. The results of this study can shed light on the observed therapeutic effects of *Urtica* on rheumatoid diseases. Caffeoylmalic acid isolated from an *U. dioica* extract and a commercial preparation of nettle leaf (IDS 23) were compared. IDS 23 showed partial inhibition of 5-lipoxygenase; isolated caffeoylmalic acid inhibited the synthesis of the leukotriene B_4 in a dose dependent manner.

Caffeoyl malic acid, a phenolic component, showed different enzymatic activities compared with IDS 23. Thus Obertreis *et al.* (1996) concluded that caffeoylmalic acid was not the only active ingredient in this commercial extract.

The activation of transcription factor NF-kappa β is elevated in several chronic inflammatory diseases. Rhiehemann *et al.* (1999) demonstrated treatment of different cells with *U. dioica* standardized preparation (IDS 23) inhibits NF-kappa β activation. They suggest that anti-inflammatory effect of *Urtica* extracts may be related to its inhibitory effect on NF-kappa β activation.

Taucher *et al.* (1996) also studied the same preparation from *U. dioica* extract (IDS 23). Lipopolysaccharide (LPS) stimulated levels of TNFα, interleukin (IL-1β) and interleukin (IL-6). IDS 23 was found to significantly reduce TNFα, IL-1β and IL-6 concentrations after LPS stimulation of these cytokines in blood.

Antonopoulou *et al.* (1996) separated total lipids of *U. dioica* leaves and identified four main classes of phospholipids as platelet activating factors (PAF) in the rabbit. The urtication caused could be due to PAF. Chrubasik *et al.* (1997) carried out a randomized *in vivo* study comparing the effect of 50 mg diclofenac plus 50 g stewed Herba *Urtica dioica* (*U. dioica* leaves) with 200 mg diclofenac. The authors concluded that stewed Herba *Urticae dioica* may enhance the anti-rheumatic effectiveness of NSAIDs.

Kavalali and Tuncel (1997) found an anti-inflammatory activity in *U. pilulifera* seed extract. This activity was measured using carrageenan induced rat paw edema test in rats. The extract showed a prolonged anti-inflammatory activity comparable to the pharmacological efficacy of indomethacin.

Urtica dioica extract is a traditionally used adjuvant in rheumatoid arthritis (Obertreis *et al.*, 1996). Kavalali and Tuncel (1999) also examined the effects of *U. pilulifera* extract on adjuvant induced arthritis in rats.

Anti-prostatic hyperplasia activity

Urtica dioica root extracts significantly inhibit prostate growth and benign prostatic hyperplasia (BPH). The parts of the plant above the ground have no effect on the prostate but have a diuretic action.

Lignans and their metabolites isolated from *U. dioica* reduce the binding activity of human SHBG (Hryb *et al.*, 1995). They also inhibit human leukocyte elastase (HLE) activity which is a clinically important marker (Barnes *et al.*, 2002).

Wagner *et al.* (1994, 1995) suggested that UDA and *Urtica* polysaccharides may also play a major role in anti-prostatic activity. Hartmann *et al.* (1996) examined the possible effects of the extract PY 102 of *Pygeum africanum* and UR 102 of *U. dioica* on the prostate gland, which are both used to treat prostate disorders. The combination of the two extracts named 'prostatonin' is made and marketed by Pharmaton of Lugano, Switzerland. Research conducted by this company showed that *Urtica* root extract was effective only at high concentration while *Pygeum* bark extract had a much higher efficacy at lower doses. Both extracts were effective in inhibiting two enzymes, 5α-reductase and aromatase. The combination of the two plant extracts, 'prostatonin', was more effective than either extract used alone in blocking aromatase activity. This showed the importance of using both extracts in the treatment of BPH.

Ali-Shtayeh *et al.* (2000) evaluated the plants used to treat skin diseases and prostate cancer in Palestine. Fifty-nine plants were reported to be effective against cancer and prostate disorders, including *U. pilulifera*.

Odenthal (1996) reported that phytosterols (constituents of *Cucurbita*, *Hypoxis*, *Pygeum*, *Urtica* and *Sabal*) are probably active against BPH. Phytoderivatives obtained from *Serenoa repens*, *Pygeum africanum*, *Cucurbita pepo* and *Urtica dioica* plant extracts are all beneficial in the treatment of BPH. Marandola *et al.* (1997) and also Bombardelli and Morazzoni (1997) reported hydroalcholic extract of the *U. dioica* roots was currently being used in the therapy of slight and moderate BPH.

Deitehoff *et al.* (2000) showed that in several clinical investigations, *Urtica* root extract improved urinary flow and reduced residual urine and prostate volume. In conclusion, evidence suggests that *Urtica* polysaccharides and also the N-acetyl-glucosamine specific lectin UDA play a major role in the anti-prostatic activity of the root extract and phytopreparations containing it.

Studies for other activities

In traditional therapy, *U. dioica* is known to have a hypotensive effect (Garnier *et al.*, 1961). Tahri *et al.* (2000) found an acute hypotensive effect after continuous perfuration of aqueous extract of aerial parts of *U. dioica* in the rat. This indicates direct effects on the cardiovascular system. Diuretic and natriuretic effects were also observed in rats by the same authors. Testai *et al*. (2002) found a possible direct cardiovascular action of *U. dioica* root extract. Anti-helminthic activity was also found in *U. dioica* roots (Manandhar, 1995).

In order to clarify the mechanism of urtication (Black and Greaves, 1991), after contact with *Urtica*; *U. dioica* and *U. urens*, hairs and whole plant extracts were examined by Czarnetzki *et al*. (1990), who found high levels of immuno-active leukotrienes (LTB$_4$ and LTC$_4$).

Bohm and Maier (1975) demonstrated the intra-vital and post-mortal mechanism of urticarial weal after contact with *Urtica* stings in albino rats. Oliver *et al*. (1991) found that the stinging sensation of the *Urtica* stings might suggest the presence of histamine in nettle fluid which is directly toxic to nerves or secondary release of other mediators.

In folk medicine *Urtica* leaf and seed has been used as a decoction for internal haemorrhages (Watt and Brandwijk, 1962). Kavalali and Akcasu (1983, 1986) found that *U. dioica* leaf extract prolongs thrombin time (TT), activated partial thrombin time (αPTT) and prothrombin time (PT) in a concentration dependent manner. The same activity was also found by Kavalali (1984) in *U. pilulifera* seed extract.

Oral and parenteral administration of *U. dioica* plant extract lowers blood sugar levels (Haznagy, 1943; Petlevski *et al*., 2001). Roman Ramos *et al*. (1992) investigated the hypoglycemic effect of 12 'antidiabetic' plants used in Mexico using 27 healthy rabbits. Blood glucose was determined every 60 minutes for a period of 5 hours. Results showed that *U. dioica* increased glycemia slightly.

Swanston-Flatt *et al*. (1989) studied 12 plants used for the traditional treatment of diabetes in northern Europe using normal and streptozotocin administered diabetic mice. Nettle plant was found to aggravate the diabetic condition in streptozotocin administered mice. Neef *et al*. (1995) found a hypoglycemic effect of nettle water extract in mice given an oral glucose load. Kavalali *et al*. (2000; 2003) found a hypoglycemic activity of the isolated lectin from *U. pilulifera* seeds in the diabetic rat models.

Medicinal plants used in folk medicine in Morocco to treat arterial hypertension and diabetes were evaluated (Ziyyat *et al*., 1997). Eighteen plant species were examined for hypertension and 14 species for diabetes. Among all the species studied, *U. dioica* was found as a remedy for hypertension and diabetes. The authors also found a close relationship between hypertension and diabetes in this region of Morocco.

The chitin binding lectin in the rhizomes and immature seeds of *U. dioica* (UDA) shows antifungal action that differs from the action of chitinase (Broekaert *et al*., 1989). *Urtica* lectin acts synergistically with chitinase in inhibiting fungal growth. Lerner and Reikhel (1992) also investigated the relationship between *in vitro* antifungal and insecticidal activities of UDA and found that it inhibited the growth of several phytopathogenic and saprophytic chitin-containing fungi *in vitro*. Cole (1994) found insecticidal activity in UDA.

Kavalali and Diren (1992) investigated the antimicrobial activity of the seeds of *U. pilulifera*. The seed extract was found to be active against *Streptococcus faecalis*, *S. aureus*, *Pseudomonas aeruginosa*, *Candida albicans* and *Escherichia coli* in a concentration dependent manner. Ethanolic extract of *U. dioica* (aerial parts) was tested for antibacterial action against seven different bacteria and found active in the minimum inhibitory concentration (Keles *et al*., 2001).

Mittman (1990) found an anti-allergic rhinitis activity in the freeze-dried preparation of *U. dioica*; Kruk and Moskalyk (2001) recommended the use of *U. dioica* juice all year round for the treatment of allergic rhinitis. Balzarini *et al.* (1992) found an anti-HIV activity of UDA *in vitro*. Dashinamzhilov *et al.* (1997) investigated *U. dioica* extract for hepatoprotective properties. The main constituent was found to be polyphytochol.

Kalaycioğlu *et al.* (1997) reported anti-mutagenic effects of an *U. dioica* extract against pesticides in the *Salmonella typhimurium*. Karakaya and Nehir El (1999) found anti-mutagenic effect of *U. dioica in vitro*.

Başaran *et al.* (1997) found immunomodulatory activities of some Turkish medicinal plants and *U. dioica* seeds and leaves were suggested to be useful for patients suffering from neutrophil function deficiency. Analgesic properties of *U. dioica* have been evaluated by classical hot plate test. No analgesic effect was noted (Tita *et al.*, 1993). Lasheras *et al.* (1986) found local anaesthetic activity of *U. dioica* extract on rat tails which was almost as powerful as lidocaine.

Among the many plant agglutinins, evaluated for antiviral activity *in vitro*, UDA was found as a potent and selective inhibitor of human immunodeficiency virus (HIV-1 and HIV-2), cytomegalovirus (CMV) and respiratory syncytial virus (RSV) replication *in vitro* (Balzarini *et al.*, 1992). Inhibitory effects of UDA on the production of gelatinase A (MMP-2) and gelatinase B (MMP-9) by mononuclear white blood cells were investigated (Dubois *et al.*, 1998). It was found to be a potent inducer of gelatinase B.

The extract of *U. dioica* was reported to be active against leukaemic cell in mice (Ilarionova, 1992). Kavalali *et al.* (1995) examined the antitumoural activity of *U. pilulifera*. The activity of the extracts was determined by the probit analysis method described by Finney (1971). In this study, the following data were found: 'Seeds ED50:12.09 and confidence upper limit: 21.4, lower limit 6.81 and Roots ED50:11.98 and confidence upper limit: 27.26, lower limit 5.12.'

Wetherilt (1989) investigated the *in vivo* and *in vitro* activities of *U. pilulifera* seeds on myeloma cells and neoplasms. The water extract of the seeds had a high activity (85–95%) in myeloma cell culture. After the purification of the product mixture showed 100% activity at 0.1 mg/ml concentration in the cell culture.

References

Adamski, R. and Bieganska, J. (1980) Studies of chemical substances present in *Urtica dioica* leaves. Part I Trace element. *Herba Pol.* 26(3): 177–180.

Ali-Shtayeh, M. S., Yaniv Z. and Mahajna, J. (2000) Ethnobotanical survery in the Palestinian area: a classification of the healing potential of medicinal plants. *J. Ethnopharmacology* 731: 221–232.

Andersen, S. and Wold J. K. (1978) Water-soluble glycoprotein from *Urtica dioica* leaves. *Phytochemistry* 17: 1875–1877.

Anikina, E. V. (1996) Physicochemical characteristics of extracts from some medicinal plant species as food additives. *Rastitel'nye Resursy* 32(4): 30–36.

Antonopoulou, S., Demopoulos, C. A. and Andrikopoulos, N. K. (1996) Lipid separation from *Urtica dioica*: existence of platelet-activating factor. *J. Agri. Food Chem.* 44(10): 3052–3056.

Balzarini, J., Neyts, J., Schols, D., Hosoya, M., Van Damme, E., Peumans, W. J. and Clercq, E. D. (1992) The mannose-specific plant lectins from *Cymbidium* hybrid and *Epipactis helleborine* and the (*N*-acetylglucosamine)n-specific plant lectin from *Urtica dioica* are potent and selective inhibitors of human immunodeficiency virus and cytohemalovirus reptication *in vitro*. *Antiviral Research* 18: 191–207.

Barlow, R. and Dixon, R. O. D. (1973) Cholin Acetyltransferase in the Nettle *Urtica dioica* L. *Biochem. J.* 132: 15–18.

Barnes, J., Anderson, L. A. and Phillipson, J. D. (2002) *Herbal Medicines, A Guide for Healthcare Professionals*, second edn, Pharmacentical Press, London.

Başaran, A. A., Ceritoğlu, I., Undeğer, U. and Başaran, N. (1997) Immunomodulatary activities of some Turkish Medicinal plants. *Phytotherapy Research* 11(8): 609–611.

Black, A. K. and Greaves, M. V. (1991) Contact Urticarie due to Common stinging Nettle Histological Ultrastructural and Pharmacological studies. *Clin. Exp. Dermatol.* 16(1): 1–7.

Beintema, J. J. and Peumans, W. J. (1992) The primary structure of stinging nettle (*Urtica dioica* agglutinin. *FEBS*. 299(2): 131–134.

Bohm, E. and Maier, R. D. (1975) Weal-formation by *Urtica dioica*, An Intravital reaction. *Z. Rechtsmed.* 5, 76(1): 1–9.

Bombardelli, E. and Morazzoni, P., (1997) *Urtica dioica* L. *Fitoterapia* 68(5): 387–402.

Booth, V. H. and Bradford, M. P. (1963) Tocopherol contents of vegetables and fruits. *Br. J. Nutr.* 17: 575–581.

Broekaert, W. F., Parijs, J. V., Leyns, F., Joos, H. and Peumans, W. J. (1989) A Chitin-binding lectin from stinging nettle rhizomes with antifungal properties. *Science* 245: 1100–1102.

Budzianowski J.(1991) Caffeic acid esters from *Urtica dioica* and *Urtica urens*. *Planta Medica*, 57, p. 507.

Chaurasia, N. and Wichtl M. (1986a) Phenylpropane und lignane aus der wurzel von *Urtica dioica* L. *Dtsch Apoth Ztg.* 126: 1559–1563.

Chaurasia, N. and Wichtl, M. (1986b) Scopoletin, 3-β-Sitosterin und Sitosterin-3-β-D glucosid aus Brennessel wurzel (*Urtica radix*). *Dtsch Apoth* Ztg. 126: 81–83.

Chaurasia, N. (1987) Phytochemische untersuchungen einiger *Urtica*-Arten unter besonderer Berücksichtigung der Inhaltsstoffe von Radix und Flores *Urtica dioica* L. Philipps-Universitat Marburg/Lahn.

Chaurasia, N. and Wichtl, M. (1987a) Flavonolglykoside aus *Urtica dioica*. *Planta Medica*, 432–434.

Chaurasia, N. and Wichtl, M. (1987b) Sterols and steryl glycosides from *Urtica dioica*. *J. Nat Prod.* 50: 881–885.

Chrubasik, S., Enderlein, W., Bauer, R. and Grabner, W. (1997) Evidence for antirheumatic effectiveness of Herbal *Urticae diocae* in acute arthritis: a pilot study. *Phytomedicine* 4(2): 105–108.

Cole, R. A. (1994) Isolation of a chitin-binding lectin with insecticidal activity in chemically defined synthetic diets. *Entomologica Experimantalis et Applicate* 72(2): 181–187.

Collier, H. O. J. and Chesher, G. B. (1956) Identification of 5-hydroxytryptamine in the sting of the Nettle (*Urtica dioica*). *Brit. J. Pharmacol.* 11: 186.

Czarnetzki, B. M., Thiele, T. and Rosenbach, T. (1990) Immunoreactive leukotrienes in nettle plants (*Urtica urens*). *Int. Arch. Allergy Appl. Immunol.* 91(1): 45–46.

Daniel, M. and Wichtl, M. (1991) A comparative examination of rhizomes from *Urtica kioviensis* and *Urtica dioica*. *Planta Medica*, 57 Sub.2.

Dashinamzhilov, Z. H. B., Azhunova, T. A., Lonshakova, K. S. and Nikolaev, S. M. (1997) The effect of the complex plant preparation polyphytochol on the course of alcoholic hepatitis. *Rastitel'nye Resursy* 33(2): 64–68.

Deitehoff, P., Lange, P. and Petrowicz, O. (2000) Therapeutics profile of Nettle Root Extract in BPH. GA/ISE Congress in Zurich-Switzerland. Abs. no: P4B/06: 3–7, 2000 ETH Zurich-Switzerland.

Delcourt, M., Peumans, W. J., Wagner, M. C. and Truffa-Bachi, P. (1996) Vβ-specific deletion of mature thymocytes induced by the plant superantigen *Urtica dioica* Agglutinin. *Cellular Immunology* 168: 158–164.

Dubois, B., Peumans, W. J., Van Damme, E. J., Van Damme, J. and Opdenakker, G. (1998) Regulation of gelatinase lectins B (MMP-9) interleukocytes by plant lectins. *Febs Lett.* 8 May, 427(2): 275–278.

Emmelin, N. and Feldberg, W. (1947) The mechanism of the common nettle (*Urtica urens*). *J. Physiol.* 106: 440–455.

Finney, D. J. (1971) *Probit Analysis*, third edn, Cambridge Univ. Press, Cambridge.

Flury, F. (1927) *Z. ges exp. Med* 56,402, in Watt, J. M. and Breyer-Brandwijk, M. G. (eds) (1962), *Medicinal and Poisonous Plant of Southern and Eastern Africa*, E&S Livingstone Ltd, Edinburg, London.

Fusseder, A., Wagner, B. and Beck, E. (1988) *Bot. Acta* 101: 211–219.

Galelli, A. and Truffa-Bachi, P. (1993) *Urtica dioica* Agglutinin. *J. Immunology* 151: 1821–1831.

Galelli, A. Delcourt, M. Wagner, M. C.,Peumans, W. and Truffa-Bachi, P. (1995) Selective expansion followed by profound deletion of mature V beta 8.3+T cells *in vivo* after exposure to the superantigenic lectin *Urtica dioica* agglutinin. *J. Immunol.* 15.154(6): 2600–2611.

Gansser, D. and Spiteller, G. (1995a) Aromatase inhibitors from *Urtica dioica* roots. *Planta Medica* 61(2): 138–140.

Gansser, D. and Spiteller, G. (1995b) Plant constituents interfering with human sex hormone-binding globulin. Evaluation of a test method and its application to *Urtica dioica* root extract. *Z.Naturforsch* 50(1–2): 98–104.

Garnier, G., Bezanger-Beanquesne and Debraux, G. (1961) Resources medicinales de la flore Française. *Vigot Freres*, Paris 2: 962–964.

Gravel, I. V., Yakovlev, G. P., Petrow, N. V., Stulovski, S. S. and Listov, S. A. (1994) Content of heavy metals in some species of medicinical plants in the Altaiskii Krai. *Rastitel'nye Resursy* 30(1–2): 101–108.

Hartmann, R. W., Mark, M. and Soldati, F. (1996) Inhibition of 5-alpha-reductase and aromatase by PHL-0081, a combination of PY 102 (*Pygeum africanum*) and UR 102 (*Urtica dioica*) extracts. *Phytomedicine* 3(2): 121–128.

Haznagy, A. (1943) Ber ungar Pharm Ges. 19 247; Chem Zbl 11, 1977/8, in J. M. Watt and M. G. Breyer Brandwijk (eds), *Medicinal and Poisonous Plants of Southern and Eastern Africa*, second edn (1962), E&S Livingstone Ltd, London.

Hegnauer, R. (1973) Chemotaxonomie der Pflanzen Band 6, Birkhauser Verlag Basel und Stuttgart. p. 629–635.

Hryb, D. J., Khan, M. S., Romas, N. A. and Rosner, W. (1995) The effect of extracts of the roots of the stinging nettle (*Urtica dioica*) on the interactions of SHBG with its receptor on human prostatic membrane. *Planta Medica* 61(1): 31–32.

Huesing, J., Murdock, L. L. and Shade, R.E. (1991) Rice and stinging nettle lectins: insecticidal activity similar to wheat germ agglutinin. *Phytochemistry* 30(11): 3565–3568.

Hughes, R. E., Ellery, T., Harry, V., Jankins and Jones, E. (1980) The dietary potential of the common nettle. *J. Sci. Food Agric.* 31: 1279–1285.

Ilarionova, M. (1992) Cytotoxic effects on leukemic cells of the essential oils from rosemary, wild geranium and nettle and concrete royal bulgarian rose. *Anticancer Res.* 12: 1915.

Kalaycioğlu, A., Oner, C. and Erdem, G. (1997) Observation of the antimutagenic potencies of plant extracts against pesticides in the *Salmonella typhimurium* strains TA98; and TA 100. *Turkish J. Botany* 21(3): 127–130.

Karakaya, S. and Nehir El, S. (1999) Quercetin, luteolin, apigenin and kaempferol contents of some foods. *Food Chem.* 66: 289–292.

Kavalali, G. (1984) Une etude sur l'activite anticoagulante d'Urtica pilulifera, XVIIIeme Semaine Medicale Balkanique, Istanbul, Turkey, Abs. no. 447.

Kavalali, G. and Akcasu, A. (1983) *Urtica dioica* L. bitkisinin Antikoagulan Etkisi, *Turkish J. Pharmacol Clin Res.* 1(2): 103–110.

Kavalali, G. and Akcasu, A. (1986) A study on the anticoagulant substance isolated from the leaves of *Urtica dioica* Thrombosis and Hemorrhagic Diseases proceedings of IVth Int. Meeting of Danubian League Against Thrombosis and Haemorrhagic Diseases. Sep. 25–28, Istanbul, Turkey, p. 244.

Kavalali, G. and Diren, Ş. (1992) *Urtica pilulifera* L. Meyvalarinin Antimikrobik Etkileri. Prof. Dr. Sarim H. Çelebioğlu anisina bilimsel toplanti. *Ist Ecz. Fak.* İstanbul May 18, Turkey.

Kavalali, G. and Tuncel, H. (1997) Anti-inflammatory Activities of *Urtica pilulifera. Int. J. Pharmacognosy* 35(2): 138–140.

Kavalali, G. and Tuncel, H. (1999) Anti-inflammatory activity of *Urtica pilulifera* seeds in adjuvant-induced arthritis in Rats. 2000 years of Natural Products Research (past, present and future), July 26–30; Amsterdam, The Netherlands, Abs. no. 778.

Kavalali, G., Dandik, L., Tuncel, H. and Aksoy, A. H. (1993) *Urtica pilulifera* L. bitkisinin tohum yaği ve Anti-enflamatuvar etkisi üzerinde bir çalişma, Xth Meeting on Vegetable Crude Drugs, Ege University, İzmir, Turkey, Abs. no. PB-34.

Kavalali, G., Tamer, L., Yurtsever, E. and İspir, T. (1995) Brine Shrimp (Artemia salina) Bioassay Results of *Urtica pilulifera* Int. Cong of Pharmaceutical Sciences (FIP 55th) Stockholm, Sweden, Abs. no. 360.

Kavalali, G., Tuncel H., Goksel S. and Hatemi, H. H. (2000) Hypoglycemic activity of *Urtica Pilulifera* in Streptozotocin diabetic rats. Natural products research in the new millennium GA/ISE congress in Zurich, Switzerland Abs. no. PZA/46 (*J. of Ethnopharmacology* 2003, 84, 241–245).

Keles, O., Bakirel, T., Ak, S. and Alpmar, A. (2001) The antibacterial activity of same plants used for medicinal purposes against pathogens of veterinary importance. *Folia veterinaria* 45: 22–25.

Korchagina, L. N., Rudyuk, V. F. and Chernobal, V. T. (1973) Lipase activity of seeds and vegetative organs of several plants. *Rast. Resur.* 9(4): 577–581.

Kraus, R. and Spiteller, G. (1990) Phenolic compounds from roots of *Urtica dioica. Phytochemistry* 29(5): 1653–1659.

Kraus, R. and Spiteller, G. (1991a) Terpene diols and Terpene diol glucosides from roots of *Urtica dioica. Phytochemistry* 30(4): 1203–1205.

Kraus, R. and Spiteller, G. (1991b) Ceramides from *Urtica dioica* roots. *Liebigs Ann. Chem.* 125–128.

Kruk, M. B. and Moskalyk, O. Ye (2001) Use of *Urtica dioica* juice in complex treatment of Allergic Rhinitis Complicated by Polypous Ethmoiditis. World Conference of Medicinal and Aromatics Plants, Hungary, Abs. no. 287.

Lasheras, B., Turillas, P. and Cenarruzabeitia, E. (1986) Étude pharmacologique préliminaire de *Prunus spinosa* L, *Amelanchier ovalis* Medikus, *Juniperus communis* Let *Urtica dioica* L. *Plant Méd. Phytothér.* 20: 219–226.

Le Moal, M. A. and Truffa-Bachi, P. (1988) *Urtica dioica* agglutinin, a new mitogen for murine T lymphocytes, unaltered Interleukin-1 production but late interleukin 2-mediated proliferation. *Cellular Immunology* 115: 24–35.

Le Moal, M. A., Colle, J. H., Galelli, A. and Truffa Bachi, P. (1992) Mouse T-lymphocyte activation by *Urtica dioica* agglutinin. *Res. Immunol.* 143: 691–700.

Lerner, D. R. and Raikhel, N. V. (1992) The gene for stinging nettle lectin (*Urtica dioica* agglutinin) encodes both a lectin and a chitinose. *J. Biol. Chem.* 267(16): 11085–11091.

Mac Farlane, W. V. (1963) The stinging properties of Laportea. *Econ. Botany* 17: 303–311.

Maitai, C. K., Talalaj, S., Njoroge, D. and Wamugunda, R. (1980) Effect of extract of hairs from the Herb '*Urtica massaica*' on smooth muscle. *Toxicon.* 18: 225–229.

Majerus, P. W. and Brodie, G. N. (1972) The binding of phytohemagglutinins to human platelet plasma membranes. *J. Biol. Chem.* 247(13): 4253–4257.

Manadhar, N. P. (1995) Medicinal folk-lore about the plants used as antihelminthic agents in Nepal. *Fitoterapia* 66(2): 149–154.

Marandola, P., Jallous, H., Bombardelli, E. and Morazzoni, P. (1997) Phytoderivatives in the management of BPH. *Fitoterapia* 68(3): 195–204.

Mikael, A., Le Moal, M. A. and Truffa-Bachi, P. (1988) *Urtica dioica* agglutinin, a new mitogen for murine T lymphocytes. *Cellular Immunology* 115: 24–35.

Mittman, P. (1990) Randomized, double blind study of freeze-dried *Urtica dioica* in the treatment of allergic rhinitis. *Planta Medica* 56: 44–47.

Neef, H., Declercq, P. and Laekeman, G. (1995) Hypoglycaemic activity of selected European plants. *Phytotherapy Res.* 9: 45–48.

Obertreis, B., Ruttkowski, T., Teucher, T., Behnke, B. and Schmitz, H. (1995) *Ex-vivo in-vitro* inhibition of lipopolysaccharide stimulated tumor necrosis factor-α and interleukin-1β secretion in human whole blood by Extarctum *Urticae dioicae* folirum. *Arzneim-Forch/Drug Res.* 46(4): 389–394.

Obertreis, B., Giller, K., Teucher, T., Behnke, B. and Schmitz, H. (1996) Antiphylogistische Effekte von Extractum *Urtica dioica* foliorum im vergleich zu Kaffeoylapfel saure. *Arzneimittel Forschung Drug Res.* 46(1): 52–56.

Odenthal, K. P. (1996) Phytotherapy of benign prostatic hyperplasia (BPH) with *Cucurbita*, *Hypoxis*, *Pygeum*, *Urtica* and *Sabal serrulate*. *Phytotherapy Res.* 10 (Supplement 1): 141–143.

Oliver, F., Amon, E. U., Breathnach, A., Francis, D. M., Saratchandra, P., Black, A. K. and Greaves, M. V. (1991) Contact Urticarie due to common stinging nettle histological ultra structural and pharmacological studies. *Clin. Exp. Dermatol.* 16(1): 1–7.

Petlevski, R., Hadzija, M., Slijepcevic, M. and Juretic, D. (2001) Effect of anti-diabetic herbal preparation on serum glucose and fructosamine in non-obese diabetic mice. *J. Ethnopharmacology* 751: 181–184.

Peumans, W. J., Ley, M. D. and Broekaert, W. F. (1984) An unusal lectin from stinging nettle (*Urtica dioica*) rhizomes. *FEBS* 177(1): 99–103.

Peumans, W. J. and Van Damme, E. J. M. (1993) Plant lectins: storage proteins with a defensive role, in Basu, J., Kundu, M. and Chekrabarti, P. (eds) *Lectins Biology, Biochemistry, Clinical Biochemistry*, Wilet Eastern, Vol. 9. pp. 27–34.

Piekos, R. and Paslawska, S. (1976) Studies on the optimum conditions of extractions of silicon species from plants with water. *Planta medica* 30: 331–336.

Radunz, A. (1976) Localization of the tri- and digalactosyl diglyceride in the thylakoid membrane with serological methods. *Z. Naturforsch, C. Biosc.* 31C (9–10): 589–593.

Regula, I. (1972) Kromatografska identifikacija alkoloide bufotenina uliutoj koprivi (*Urtica pilulifera* L.) *Acta bot. Croat* 31: 109–112 C.A. 74, 95464 in Hegnauer, R. (1973) *Chemotaxonomie der pflanzen Band* 6., p. 635.

Riehemann, K., Behnke, B. and Schulze-Osthoff, K. (1999) Plant extract from stinging nettle (*Urtica dioica*) and antirheumatic remedy, inhibit the proinflammatory transcription factor NF-$\kappa\beta$. *FEBS letters* 442(1): 89–94.

Roman-Ramos, R., Alarcon-Aguilar, F., Lara-Lemus, A. and FloresSaenz, J. L. (1992) Hypoglycemic effect of plants used in Mexico as antidiabetics. *ArchMed-Res.* **23**(1): 59–64.

Roslon, W. and Weglarz, Z. (2001) Polyphenolic acid of female and male forms of nettle (*Urtica dioica*). World Conference on Medicinal and Aromatic plants, Hungary, Abs. no. 186.

Rovira, P., Buckle, M., Abastado, J. P., Peumans, W. J. and Truffa-Bachi, P. (1999) Major histo-compatibility class I molecules present *Urtica dioica* agglutinin, a superantigen of vegetal origin, to T lymphocytes. *Eur. J. Immunology* **29**(5): 1571–1580.

Rzemykowska, Z. and Ostrowska, B. (1994) The method of quantitative determination of magnesium in the juice from the fresh nettle (*Urtica dioica*). *Herba Polonica*, **40**(3): 95–98.

Salamon, I., Hecl, J. and Haban, M. (2001) Heavy metal determination of several medicinal plant in the central Zemplin. World Conference on Medicinal and Aromatic plants, Hungary, Abs. no. 187.

Saxena, P. R., Pant, M. C., Kishor, K. and Bhargava, K. P. (1965) Identification of pharmaco-logically active substances in the Indian stinging Nettle, *Urtica parviflora* Roxb. *Canad. J. Pharmacol.* **43**: 869–876.

Schilcher, V. H. and Effenberger, S. (1986) Scopoletin und β sitosterol-zwei geeignete Leitsubstanzenfur *Urtica radix. Deutsche Apotheker Zeitung* **126**(3): 79.

Schomaker, V. J., Bollbach, F. D. and Hagels, H. (1995) Brennesselkraut. *Deutsche Apotheker Zeitung* **135**(7): 40–45.

Schottner, M., Gansser, D. and Spiteller, G. (1997) Lignans from the roots of *Urtica dioica* and their metabolites bind to human sex hormone binding globulin (SHBG). *Planta Medica* **63**(6): 529–532.

Shibuya, N., Goldstein, I. J., Shafer, J. A., Peumans, W. J. and Broekaert, W. F. (1986) Carbohydrate binding properties of the stinging nettle (*Urtica dioica*) rhizome lectin. *Arch. Biochem. Biophy.* **249**(1): 215–224.

Smallman, B. N. and Maneckjee, A. (1981) The synthesis of acetylcholine by plants. *Biochem. J.* **194**: 361–364.

Stillmark, H. (1988) Thesis, University of Dorpat (Tartu).

Swanston-Flatt, S. K., Day, C., Flatt, P. R., Gould, B. J. and Bailey, C. J. (1989) Glycaemic effect of traditional European plant treatment for diabetes, Studies in normal and streptozotocin diabetic mice. *Diabetes Res.* **10**(2): 69–73.

Tahri, A., Yamani, S., Leggsyer, A., Aziz, M., Mekhfi, H., Bnouham, M. and Ziyyat, A. (2000) Acute diuretic, natriuretic and hypotensive effects of a continuous perfusion of aqueous extract of *Urtica dioica* in the rat. *J. Ethnopharmacology* **732**: 95–100.

Taucher, T., Obertreis, B., Ruttkowski, T. and Schmitz, H. (1996) Cytokine secretion in whole blood of healthy volunteers after oral ingestion of an *Urtica dioica* leaf extract. *Arzneimttel Forschung* **46**(8): 906–910.

Testai, L., Chericoni, S., Calderone, V., Nencioni, G., Nieri, P., Morelli, I. and Martinotti, E. (2002) Cardiovascular effects of *Urtica dioica* L. (*Urticaceae*) roots extracts: *in vitro* and *in vivo* pharmacological studies. *J. Ethnopharmacology* **81**: 105–109.

Tita, B., Faccendini, P., Bello, U., Martinoli, L. and Bolle, P. (1993) *Urtica dioica*: pharmacolog-ical effect of ethanol extract. *Pharmacol. Res.* **27**(1): 21–22.

Ullrich, I. and Jahn-Deesbach, W. (1984) Proteingehall und proteinzusammensetzung verschidener unkrautarten. *Angrew. Bot.* **58**(3–4): 255–266.

Van Damme, E. J. M. and Peumans, W. J. (1987) Isolectin composition of individual clones of *Urtica dioica*: evidence for phenototypic difference. *Physiol. Plant* **71**: 328–334.

Van Damme, E. J. M., Broekaert, W. F. and Peumans, W. J. (1988) The *Urtica dioica* agglutinin is a complex mixture of isolectins. *Plant Physiol.* **86**: 598–601.

Van Damme, E. J. M.,Goldstein, I. J. and Peumans, W. J. (1991) A comparative study of mannose binding lectins from the Amaryllidaceae and Alliaceae. *Phytochemistry* **30**(2): 509–514.

Vishnevskii, O. V. and Proshunina, D. V. (1989) Farm Zh. 2, 50–53, in *Food Science and Human Nutrition*, G. Charalambus (ed.), Elsevier Science, Published in 1992.

Wagner, H. and Willer, F. (1990) Chemie und Pharmakologie von *Urtica*-praparaten. *Urologie* **3**: 309–312.

Wagner, H. and Willer, F. (1992) Zur Chemie und Pharmakologie der Polysaccaharide und Lectine von *Urtica dioica* Wurzeln, in G. Rutishauser (ed.), *Benign Prostatahyperplasie III. Klein. Exp. Urol.* **22**: 125–132 W. Zuckschwerdt-Verlag.

Wagner, H., Willer, F. and Kreher, B. (1989) Biologisch aktive Verbindungen ans dem Wasserextrakt von *Urtica dioica*. *Planta Medica* **55**: 452–453.

Wagner, H., Willer, R., Samtleben, R. and Boos, G. (1994) Search for the antiprostatic principle of stinging nettle (*Urtica dioica*) roots. *Phytomedicine* **1**: 213–224.

Wagner, H., Geiger, W. N., Boos, G. and Samtleben, R. (1995) Studied on the binding of *Urtica dioica* agglutinin (UDA) and other lectins in an *in vitro* epidermal growth factor receptor test. *Phytomedicine* **4**: 287–290.

Watt, J. M. and Breyer-Brandwijk, M. G. (1962) *Medicinal and Poisonous Plants of Southern and Eastern Africa*, second edn, E&S Livingstone Ltd, Edinburg and London. p. 1043.

Weglarz, Z. and Roslon, W. (2000) Developmental and Chemical variability of female and male forms of nettle *Urtica dioica* L. *Acta Horticulturae* **523**: 75–80.

Wetherilt, H. (1989) Assessment of the nutritional properties and antitumoural activity of the common nettle grown in Turkey. PhD thesis, Hacettepe Univ. Ankara.

Wojtaszek, E. M., Bylka, W. and Kowalewski, Z. (1986) *Herba Polonica* **32**: 131, in Budzianowski, J. (1991) *Planta Medica* **57**, p. 507.

Ziyyat, A., Legssyer, A., Mekhfi, H., Dassouli, A., Serhrouchni, M. and Benjelloun, W. (1997) Phytotherapy of hypertension and diabetes in oriental Moroccos. *J. Ethnopharmacology* **58**: 45–54.

4   Various therapeutic uses of *Urtica*

Colin Randall

Introduction

In this chapter the prescribed treatments using *Urtica* will be discussed. Only those treatments that are supported by scientific evidence will be considered. These treatments are for arthritis and rheumatism, allergic rhinitis, and benign prostatic hypertrophy.

Mode of use of nettle

Treatment with external application of nettle leaf

Exploratory study of the use of nettle sting for musculoskeletal pain (Randall et al., 1999)

The author's exploratory survey reported the use of nettle sting by 10 males and 8 females, age range 42–82, living in the United Kingdom. All were self-prescribing and using fresh nettle leaves (*Urtica dioica*) to raise urticarial weals on areas requiring relief from musculoskeletal pain. Only the 82-year-old male, awaiting a second knee joint replacement, did not benefit. Some interviewees considered themselves cured. No adverse reports were recorded.

The findings from this study, in particular application, dosage and effect of nettle sting (see Tables 4.1 and 4.2), were used to plan a randomised controlled treatment trial.

Randomised controlled trial of nettle sting for joint pain (Randall et al., 2000)

A randomised controlled crossover treatment trial was carried out in a rheumatology department and a general practice in Plymouth, United Kingdom, during 1998. Twenty-seven patients with persistent wrist/thumb pain were studied. The pain relief effect of once daily application for one week of nettle leaf (*Urtica dioica*) was compared with white dead-nettle leaf (*Lamium album* – non-stinging plant which looks like nettle) as placebo. The results showed a statistically significant reduction in pain and disability score. Other recorded parameters of pain, namely analgesic and anti-inflammatory

Table 4.1 Method of treatment application

Application	No. of patients
Method of application	
Stroking affected area	9
Beating affected area	5
Pressing leaf on affected area	4
Time of contact with skin	
Less than 30 seconds	11
Up to 2 minutes	3
3–4 minutes	3
1–2 hours (with bandage)	1
No. of applications per treatment	
One	5
Two	2
Three	2
Four	1
Twelve	2
More than twenty	3
>50, 2–3 times a week	1
Daily for nine months	1
Daily for two years	1

Source: Randall *et al.*, 1999.

consumption levels, sleep, and patients' global assessment, all provided supporting evidence of an analgesic effect.

The apparent mode of pain relief seems to be associated with repeated production of urticarial weals. In addition, a warm tingling is frequently reported. This tingling usually follows within 30 minutes as the stinging sensation recedes, and then continues for 12–24 hours or more. A subsequent local area of anaesthesia is often reported. This stinging sensation tends to be less after repeated daily stinging while the analgesic effect is usually found to increase. The warm tingling sensation diminishes with repeated applications as the analgesia increases. Interestingly, this is similar to the experiences reported by patients using capsaicin ointment.

The author describes the most probable mechanism of nettle sting analgesia as *thermal hyper-stimulation analgesia*. Nettle sting causes a prolonged warm tingling hyper-stimulation, which lasts over 24 hours after one application. It is speculated that a profound thermal hyperalgesia provoked by serotonin and possibly other chemical constituents of nettle sting cause a hyper-stimulation of afferent pain fibres that subsequently cause an inhibitory blocking action on the dorsal horn cells of the 'gate' in the spinal cord (Melzack and Wall, 1965; Wall and Sweet, 1967; Randall, 2001).

Another possibility is that nettle sting analgesia has a similar mechanism to capsaicin analgesia, which also causes warm stimulation and increased analgesia with continued use. The mode of action is thought to be a consequence of depletion and

Table 4.2 Treatment effect

Treatment effect	No. of patients
Pain relief from nettles?	
Every time	15
90% of times	2
No pain relief	1
Time interval before pain relief	
10 minutes	4
Between 10 minutes and 6 hours	3
6–24 hours	4
36 hours–4 weeks	6
No pain relief	1
Pain relief period after treatment	
1–7 days	5
2–24 months	6
> 3 years (maximum 9)	5
Yes but no record of time period	1
None	1
Side-effects*	
No serious side-effects	18
'Not unpleasant warmth'	14
Intense pain changing to warmth	2
Area numbness for 6–24 hours	3
Rash (apart from usual weals)	3
Itchy rash, throbbing later in day	1

Source: Randall *et al.*, 1999.

Note
* Some patients reported more than one side-effect.

blockade of synthesis of substance P from the sensory nociceptors C fibre neurones (Rains and Bryson, 1995; Randall, 2001).

Nettle taken as an internal medication

Nettle leaf extract for allergic rhinitis

In the USA Dr Andrew Weill has recommended 1–2 freeze-dried nettle extract capsules every 2–4 hours for hay fever and allergic sinus problems (Weill, 1995). This use is supported by a randomised double blind study by Mittman using these capsules for allergic rhinitis. In this double-blind randomised study of 98 individuals, the effects of a freeze-dried preparation of *U. dioica* were compared with placebo on allergic rhinitis. Sixty-nine individuals completed the study. Assessment was based on daily symptom diaries, and global response recorded at the follow-up visit after one week of therapy. *Urtica dioica* was rated higher than placebo in the global assessments. Comparing the diary data, *U. dioica* was rated only slightly higher (Mittman, 1990).

Nettle leaf extract for arthritis

Nettle leaf extract in capsule form is prescribed in Germany for the pain of osteo- and rheumatoid arthritis, and two clinical treatment studies (not randomised controlled) were recently reported (Chrubasik *et al.*, 1997; Chrubasik and Eisenberg, 1999). A large study of 8,955 patients taking 1,340 mg *Urtica* extract IDS-23 per day produced encouraging results of improvement in pain at rest (55%), pain on exercise (45%), physical impairment (38%), and reduction in consumption of non-steroidal anti-inflammatories in 60% of patients. The onset of effectiveness occurred after 11 days (Chrubasik and Eisenberg, 1999).

Nettle seed treatment for rheumatism

In *Herbal Drugs and Phytopharmaceuticals: A Handbook for Practice on a Scientific Basis*, it is stated that 'nettle fruit-seeds are used principally in folk medicine but there is no scientific evidence for crushed seeds being used as a dressing for skin complaints and rheumatism' (Bissett, 1994). However, recently published scientific evidence has demonstrated a dose dependent anti-inflammatory activity of *U. pilulifera* (Roman nettle) seed extract in rats, in an experimental module (Kavalali and Tuncel, 1997). In this study the anti-inflammatory action of *U. pilulifera* seed extract is compared with indomethacin and placebo using a rat paw carrageenan-induced oedema experimental module. Results showed a significant and similar dose related anti-inflammatory action with both *U. pilulifera* seed extract and indomethacin compared with placebo.

Nettle root extract for benign prostatic hypertrophy

In a Commission E monograph (1991) for nettle root extract, the German authorities recognised its use for benign prostatic hypertrophy when it was taken as a root tea crude drug in a dose of 4–6 grams daily (Schulz *et al.*, 1997). In their comprehensive review article '*Urtica dioica*' Bombardelli and Morazzoni summarise the research evidence up to 1997 (Bombardelli and Morazzoni, 1997). The most important studies of internally taken nettle medication for benign prostatic hypertrophy are two placebo controlled double blind trials, both using a dosage of 600 mg nettle root extract daily. In these studies, one with 50 patients (Vontobel *et al.*, 1985), and the other with 79 patients (Dathe and Schmid, 1987), there was a statistically significant improvement in urine flow compared with placebo. Three large multi-centred studies were also reviewed, all of which had more than 4,000 patients. Two studies showed subjective improvement (Tosch and Mubiggang, 1983; Stahl, 1984) and the other also reported improved mean urine flow, and reduction in frequency, nocturia, and residual urine (Friesen, 1988). These clinical studies all used hydroalcoholic nettle root extracts prepared with relatively hydrophilic solvents, that is, methanol or ethanol in concentrations of 20–60%. The main components of these extracts include phytosterols, triterpene acids, lignans, polysaccharides, and simple phenol compounds (Schulz *et al.*, 1997). In all of these studies of nettle root extract for benign prostatic hypertrophy the tolerability was said to be very good (Bombardelli and Morazzoni, 1997).

In an observational study of 4,087 patients with benign prostatic hypertrophy taking 600–1,200 mg of nettle root extract daily for six months, maximum urinary flow and residual urine volume improved in 50–60% of patients but only 35 (0.86%) reported side-effects. Thirty-three cited gastro-intestinal complaints (0.65%), 9 noting skin allergies (0.19%), and 2 reporting hyperhydrosis (some patients had more than one side-effect) (Sonnenschein, 1987).

Toxicology

A search of the medical and Toxline databases have revealed no serious side-effects when humans take the usually recommended oral doses of *U. dioica*, *U. urens*, or *U. pilulifera*. Several publications have stated that contra-indications are unknown and side-effects, of mostly gastro-intestinal irritation, are rare (Bradley, 1992; Tyler, 1994).

Acute and chronic studies of nettle toxicity have been carried out in Spain in 1984. Mice were given an infusion of intravenous nettle, and the LD_{50} was found to be 1.92 g dry plant/kg body weight. When the nettle extract was obtained with boiling water the LD50 was 1.72 g/kg (Patten, 1993). There has been one isolated and unsubstantiated claim of lameness and death among dogs moving amongst nettles (McBarron, 1976). However this was in the Australian literature and probably relates to *U. urens,* which is the commonly occurring nettle in Australia (Patten, 1993).

The USA Food and Drug Administration consider nettle taken internally as a drug of undefined safety (Patten, 1993). There have been reports of nettle stimulating uterine contractions in rabbits, and it should not therefore be given internally to pregnant or nursing mothers. It is considered relatively safe for healthy non-pregnant, non-nursing adults not taking other diuretics, in the usual recommended doses (Castleman, 1991). Duke has reported the following toxic effects from drinking nettle tea: gastric irritation, burning sensation of the skin, oedema, and urine suppression (Duke, 1989).

Studies have revealed that nettle plants can absorb heavy metals (Ernest, 1987) and pesticide residues (Benecke, 1987). It is therefore important that nettle plants harvested for food or medicinal consumption are known to be free of pesticides and heavy metal contamination.

Summary

Indications

There is scientific evidence to support the medicinal use of *Urtica* for the following conditions – arthritis and rheumatism, allergic rhinitis, and benign prostatic hypertrophy.

Cautions

Nettles medication, both topical and internal, should be avoided if there is any known allergy to nettles, other than the usual localised urticarial rash with stinging. However, the author has found no reports of severe allergy or idiosyncrasy (except one recent report of possible severe widespread skin rash reaction after stinging – personal communication), but recommends discussion with and advice from the patient's medical adviser before commencing nettle therapy.

Contra-indications

Nettle therapy is contra-indicated in diabetics (Bissett, 1994). Newall warns against excessive (oral) use in all patients in view of 'documented irritant properties' (Newall *et al.*, 1996).

Side-effects

Side-effects of internal use are said to be rare but the following have been reported: gastrointestinal, allergic rashes, hypoglycaemia, hyperglycaemia, and oliguria/oedema (Bradley, 1992; Bissett, 1994). Oral nettle products have been variously reported as causing raised blood sugar (hyperglycaemia) (Oliver-Bever and Zahland, 1979) and lowering blood sugar (hypoglycaemia) (Wren, 1988; Newall *et al.*, 1996). Massive exposure to stinging has been reported to cause symptoms of shock in animals, but this has not been reported in humans (Wren, 1988).

Dosage

External application of stinging leaf (urtication)

The dose is not well established and different methods of application have been reported as successful in producing an analgesic effect (see Table 4.1 above). The most common dosage is to produce stinging (by pressing, stroking, or beating) for less than 30 seconds daily, for 1–7 days. This is repeated when (or if) the pain returns (Randall *et al.*, 1999; Randall, 2001).

Internal herb leaf extract

Osteo and rheumatoid arthritis: Urtica extract IDS-23 1,340 mg daily (Chrubasik *et al.*, 1997; Chrubasik and Eisenberg, 1999).

   Allergic rhinitis: 1–2 freeze-dried nettle extract capsules every 2–4 hours for 'hay fever and allergic sinus problems' (Weill, 1995). In the Oregon Portland study a dose of two 300 mg capsules of freeze-dried *U. dioica* was used, with a daily dosage ranging from 1 to 7 doses and a mean of 2.8 (Mittman, 1990).

Nettle root extract

Benign prostatic hypertrophy: 600–1,200 mg daily (Vontobel *et al.*, 1985; Dathe and Schmid, 1987; Sonnenschein, 1987).

References

Benecke, R. (1987) Residues of pesticides in drugs grown from wild grown medicinal plants. *Pharmazie*, 42, 869–71.
Bisset, N. G. (1994) *Herbal Drugs and Phytopharmaceutical.* Medpharm, Stuttgart.
Bombardelli, E. and Morazzoni, P. (1997) *Urtica dioica* L. Fitoterapia, 67(5), 387–402.

Bradley, P. (1992) *British Herbal Compendium*, Vol.1. British Herbal Medicine Association, Exeter UK, pp. 166–67.

Castleman, M. (1991) *The Healing Herbs*. Rodale Press, Emmaus Pennsylvania.

Chrubasik, S. and Eisenberg, E. (1999) Treatment of rheumatic pain with kampo medicine in Europe, Part 2 *Urtica dioica. Pain clinic*, 11(3), 179–85.

Chrubasik, S., Enderlein, W., Bauer, R., and Grabner, W. (1997) Evidence for antirheumatic effectiveness of stewed *Herba urticae dioicae* in acute arthritis: a pilot study. *Phytomedicine*, 4, 105.

Dathe, G. and Schmid, H. (1987) Treatment of benign prostatic hypertrophy with *Urtica dioica extract. Urologie* [B], 27, 223.

Duke, J. A. (1989) *Handbook of Medicinal Herbs*. CRC Press, Florida.

Ernest, W. H. O. (1987) Perennial herbs as monitor for moderate levels of metal fall-out. *Chemosphere*, 16, 223–38.

Friesen, A. (1988) Benigne prostatahyperplasie II. *Klin. Exp. Urol.*, 19, 121.

German commission E monograph (1991) American Botanical Council, Austin, Texas (USA).

Kavalali, G. and Tuncel, H. (1997) Anti-inflammatory activities of *Urtica pilulifera. Int. J. Pharmacognosy*, 35(2), 138–40.

McBarron, E. J. (1976) *Medical and Veterinary Aspects of Plant Poisons in New South Wales*. Department of Agriculture, New South Wales.

Melzack, R. and Wall, P. D. (1965) Pain mechanisms: a new theory. *Science*, 150, 971.

Mittman, P. (1990) Randomised, double-blind study of freeze-dried *Urtica dioica* in the treatment of allergic rhinitis. *Planta Medica*, 56(1), 44–7.

Newall, C. A., Anderson, L. A., and Phillipson, J. D. (1996) *Herbal Medicines: A Guide for Health-care Professionals*. The Pharmaceutical Press, London.

Oliver-Bever, B. and Zahland, G. R. (1979) Plants with oral hypoglycaemic activity. *Quart. J. Crude Drug Res.*, 17, 139–96.

Patten, G. (1993) Urtica. *Austr J. Med. Herbalism*, 5(1), 5–13.

Rains, C. and Bryson, H. M. (1995) Capsaicin. *Drugs and Aging*, 7(4), 317–28.

Randall, C. F. (2001) *Treatment of Musculoskeletal Pain with the Sting of the Stinging Nettle: Urtica dioica*. MD thesis. University of Plymouth, Plymouth, UK.

Randall, C. F., Meethan, K., Randall, H., and Dobbs, F. (1999) Nettle sting of *Urtica dioica* for joint pain – an exploratory study of this complementary therapy. *Complementary Therapies in Medicine*, 7, 126–31.

Randall, C. F., Randall, H., Hutton, C., Sanders, H., and Dobbs, F. (2000) Randomized controlled trial of nettle sting for treatment of base-of-thumb pain. *J. Royal Soc. Medicine*, 93, 305–9.

Schulz, V., Hansell, R., and Tyler, V. E. (1997) *Rational Phytotherapy: A Physicians' Guide to Herbal Medicine*. Springer-Verlag, Berlin and New York, pp. 228–9.

Sonnenschein, R. (1987) Untersuchung der Wirksamkeit eines prostatotropen Phytotherapeutikums (Urtica plus) bei benigner Prostatahyperplasie und Prostatitis – eine prospektive multizentrische Studie. *Urologe {B}*, 27, 232–7.

Stahl, H. P. (1984) Therapy of prostatic nocturia with standardized extract of urticae root. *Z. Allg Med.*, 60(3), 128–32.

Tosch, U. and Mubiggang, H. (1983) Medikamentose Behandlung der benignen Prostatahyperplasie. *Euromed*, 6, 334.

Tyler, V. E. (1994) *Herbs of Choice: The Therapeutic Use of Phytomedicinals*. Pharmaceutical Products Press/ Haworth Press, New York and London, pp. 84–5.

Vontobel, H. P., Herzog, R., Rutishauser, G., and Kres, H. (1985) *Urologie {A}*, 24, 49.

Wall, P. D. and Sweet, W. (1967) Temporary abolition of pain in man. *Science*, 155, 108.

Weill, A. (1995) *Natural Health, Natural Medicine*. Houghton-Mifflin, New York.

Wren, C. (1988) *Potters New Cyclopaedia of Botanical Drugs and Preparations*. Health Sciences Press, Bradford England.

5 The therapeutic uses of *Urtica* in benign prostatic hyperplasia

Johannes J. Lichius

Introduction

The use of stinging nettle (*Urtica dioica* L.) root extracts in the therapy of benign prostatic hyperplasia has a rather short tradition. The basis is a report by a German physician about the use of tea in the treatment of urinary tract disorders (Rückle, 1950).

Biochemical models

Inhibition of 5α-reductase

Testosterone is transformed by 5α-reductase into dihydrotestosterone (DHT). DHT is the active androgen in the prostate (dihydrotestosterone hypothesis). Androgen deprivation has been shown to decrease the size of the prostate. Nettle root extracts come off badly in studies of their effect on 5α-reductase. Respective studies were made of 60% ethanolic extract (Koch, 2001) and a 20% methanolic extract (Rhodes *et al.*, 1993), both of which turned out to be ineffective. The study made by Rhodes and coworkers, however, is not without certain shortcomings. For instance, it compares data in mg/ml and not on the basis of daily doses, without taking into account that the daily dose shows distinct differences between finasterid and plant preparations (finasterid 1–5 mg/day; bazoton 600 mg of 20% methanolic extract, corresponding to 6,300 mg of drug per day). In addition to inaccuracies regarding the botanical names there is no consideration of the excipient. In spite of such criticism it remains a fact that an activity of the extracts cannot be explained as a consequence of an inhibited 5α-reductase.

Inhibition of aromatase

With increasing age the balance of androgen/estrogen in the serum and in the prostate shifts in favour of estrogens. By blocking the aromatase the growth of the prostate is influenced. Clinical studies provided an impulse for investigating the effect of plant extracts on aromatase (estrogen hypothesis). It was found that during a twelve-week-treatment with 2*600 mg of 20% methanolic nettle root extract, there was a significant reduction of estradiol and estrone in the serum (Bauer *et al.*, 1988). After a bioassay-guided isolation it was found that the extract contained a weak aromatase inhibitor (9-hydroxy-10,12-octadecadienacid) (Ganßer and Spiteller, 1995; Kraus, 1990).

Possibly this substance is only a pro-drug taking effect by way of a metabolic product, the 9-oxo-compound known to be an active aromatase inhibitor. It remains an open question whether or not the quantities contained in the extract are sufficient for a therapeutic effect or if an accumulation can be assumed, or else if other mechanisms are responsible for the therapeutic activity.

Interaction with sex hormone binding globulin (SHBG)

SHBG is biosynthesized in the liver and has two binding sites, one for sexual hormones and one for membrane receptors identified in the prostate tissue and the endometrium (mucous membrane of the uterus). The hypothesis is structured as follows: (1) Only free SHBG is bound to the membrane receptors. (2) A steroid is bound to the docked SHBG. (3) This causes an increase of intracellular cAMP. Hryb *et al.* and Rosner found that a hot water extract of the nettle root (rich in polysaccharides) dose-dependently inhibits the binding of SHBG to the membrane receptors (Rosner, 1994; Hryb *et al.*, 1995). Urtica dioica agglutinin (UDA), stigmasta-4-en-3-on and a 70% ethanolic nettle root extract were not active in this system. Schöttner *et al.* was able to prove that a number of lignanes of nettle root are bound to the membrane receptors (Schöttner *et al.*, 1997). Free SHBG is very important for the regulation of the plasma concentrations of free androgens and estrogens. Bauer reports on a clinical study in which a decrease of estradiol and estrone values as well as the SHBG concentrations were observed (Bauer *et al.*, 1988). Further research will be necessary to find out to what extent the clinical picture corresponds to the *in vivo* experiment.

Inhibition of leucocyte elastase

Human leucocyte elastase (HLE) is a sensitive and quantitative marker for clinically inconspicuous infections of the genital tract. HLE is a serine endopeptidase. It has a low substrate specificity and is counted among the destructive enzymes.

The human prostate is run through by smooth muscles. In the course of Benign Prostatic Hyperplasia (BPH) there is a distinct hyalinization of the periacinal connective tissue and a hyperelastosis. Progessively these may result in a degeneration and fragmentation of the elastic fibres. Apart from the enlargement of the prostate it is, among other things, the loss of elasticity which is responsible for the obstructive symptoms. The inhibition of the proteolytic activity of HLE might be a useful contribution to the elimination of the BPH symptoms. Nettle root extract (60% ethanolic extract) turned out to be highly active, resulting in a dose-dependent inhibition of the degeneration of peptide substrates by HLE (Koch *et al.*, 1995).

Immunomodulatory activity

Hyperplastic prostate tissue has some particular characteristics: special, unique antigenic structures and lower titers of certain specific prostate antigens (as compared with a normal prostate). This is why it seems to be useful to discuss an immunological therapeutic approach. An immunomodulating activity is attributed to lectin (UDA) as well as to the polysaccharide fraction.

Thus, Wagner *et al.* were able to show that aqueous extracts and polysaccharide fractions gained from them have a stimulating effect on T-lymphocyte proliferation (*in vitro* lymphocyte transformation test) (Wagner *et al.*, 1989, 1994). This group was also able to show a stimulating effect of UDA on the lymphocyte proliferation.

Influencing complement activation

The complement system consists of about two dozen different serum proteins influencing each other like a cascade and is the most important effector system of inflammatory responses. Apart from the classical activation (immune reaction) the complement system can also be activated in an alternative way. This is why bacterial polysaccharides or proteolytic enzymes are able to activate the complement system. Polysaccharides gained from nettle root extracts were extremely effective as anti-complements both in the classical as well as the alternative approach (Willer, 1992).

Influencing growth factors

On the basis of the hypothesis of an epidermal-stromal interaction regulated by growth factors, investigations made by the team of Wagner analyzed *in vitro* the influence of UDA on receptors of growth factors (Wagner *et al.*, 1989, 1994, 1995). UDA is bound to the epidermal growth factor receptor (EGF-R) as well as to the fibroblast growth factor receptor FGF-R (Geiger *et al.*, 1996, 1997). Among the processes following the receptor binding there is an activated phytotyrosine kinase. This is inhibited by UDA.

It should be noted that the occurrence of EGF-R in the basal cells of the prostate is a definite fact; however, only a few not very conclusive investigations have been made about its function in BPH. The bFGF receptor was identified in the prostate stroma, but its role in the pathogenesis requires further research.

Interaction with Na^+,K^+-ATPase

At the membrane-binding site of androgen a Na^+,K^+-ATPase is located. By influencing the enzyme a change of the metabolic energy of the cell is effected. Excess growth could be inhibited in this way. Hirano *et al.* investigated the inhibiting influence of hydrophobic components of the nettle root on the untreated human prostate Na^+,K^+-ATPase (Hirano *et al.*, 1994). They showed that the inhibiting activity of the respective steroids (found in the nettle root, e.g. β-sitosterol) is in a concentration range of 10^{-3} to 10^{-6}.

Cell cultures

Fibroblast cultures from ventral rat prostates were used to determine the effect of nettle root extract (20% methanolic extract). By means of a solution of the extract (concentration 0.01%) the number of cells could be lowered by 50%. No proof was furnished of either dosage dependance or androgen dependance (Schmitt *et al.*, 1987). Explant samples of human prostate tissue cells were cultivated for some weeks. The morphology of the cells remained unchanged, the growth rate of the medicated group decreasing significantly as compared with the control group given no medication (Enderle-Schmitt *et al.*, 1988). Rausch *et al.* continued the studies undertaken by Enderle-Schmitt (Rausch *et al.*, 1992). They found that 20% methanolic nettle root extract inhibits the growth of explant cultures (human prostate tissue) by 20%, stigmastenon by 30–50%, and ceramide 50–80%. The 20% methanolic extract and the polysaccharide fraction of this extract were then examined in hPCPs and LNCaP cell culture lines. The cell cultures used for the experiments are representative of the organ to a certain extent; the stromal cell culture line (hPCPs) and the epithelial cell culture line (LNCaP). In these experiments a 20% methanolic extract and the polysaccharide fraction of this extract had only an inhibitory effect on the growth of epithelial cells (LNCaP) (Lichius *et al.*, 1999a, 1999b; Konrad *et al.*, 2000).

Animal experiments

Natural BPH occurs in humans, whereas in the animal kingdom it is only found in apes, lions, and dogs. Thus it does not come as a suprise that animal experiments in basic research were mainly conducted with dogs. In this context it should be noted that the pattern of ailments is different in dogs and humans. Contrary to humans, dogs usually have defecation rather than miction disorders, which are rare. Moreover there are histological differences in the form the disease takes (diffuse, glandular and, to some extent, cystic hyperplasia). Other studies deal with the physiology of the prostate in rodents and their organ preparations.

The most important experiments are summed up as follows: A pilot study conducted by Daube investigated the therapeutical effect of 20% methanolic nettle root extracts on BPH on dogs (Daube, 1988). The prostate volume of ten medicated dogs showed an average decrease of 30% during a 100-day therapy. The volume was determined sonographically. Scapanini and Friesen made a study of the antiandrogenic effect of nettle root extracts on castrated rats with an implanted testosterone donator (Scapanini and Friesen, 1992).

The implants were filled with increasing testosterone concentrations. On 11 days 20% methanolic nettle root extract was administered. By this therapy the weight increase of the prostate was significantly diminished compared with the control group. Similar experiments were made by Rhodes *et al.*, who gave castrated young rats a daily dose of testosterone or DHT-propionate (i.p.) (Rhodes *et al.*, 1993). In these experiments a normal prostate was used, whose growth was induced by hormone shifting. Simultaneously, the rats were treated with a preparation containing a 20% methanolic nettle root extract; the dose administered was 1/5 the dose per kg for humans. There was no verifiable effect on prostate growth. Why such a low dose was chosen is hard to understand, because normally ten times the dose for humans is used in experiments with rats. The antiphlogistic activity of substances can be verified in the rat's paw edema model. This measures the ability of a substance to reduce an edema caused by carragenan. The molecuar mechanism has not been completely understood, but a possible participation of lymphocytes and the complement system is under discussion. Willer was able to show that an orally applied raw fraction of polysaccharides from nettle roots achieved an inhibition of the induced growth by 36.8% (Willer, 1992). The hypothesis of embryonic re-awakening according to McNeal says that the mechanism of epithelio-stromal interaction during embryonic development, as described by Cunha and coworkers, is reactivated in BPH, the stroma providing the impulse for glandular growth (McNeal, 1978; Cunha *et al.*, 1980).

An *in vivo* model was developed that showed that the implantation of urogenital sinus tissue has a growth stimulating effect on the prostate of the mouse (Chung *et al.*, 1984). In this model, a combination of nettle root extract (60% ethanolic extract) and saw palmeto extract (90% ethanolic extract) was tested in a first series of experiments. Medication took place from day 3 to 18 or from day 30 to day 45 after the operation. However, there were no ascertainable effects on the weight and the histological structures of the prostate (Koch, 1995). The same model was used with slightly different conditions (e.g. medication from day 1 to day 28) in another study.

Five different extracts prepared with solvents of increasing polarity (cyclohexane, ethyl acetate, 1-butanol, 20% methanol and water) were evaluated. Only the methanolic extract caused a significant growth inhibition of about 51% (Lichius and Muth, 1997). Some compounds like UDA, polysaccharides, and secoisolariciresinol were considered to be active substances in this extract. In further experiments only the polysaccharide fraction of the methanolic extract showed a significant inhibiting activity (Lichius *et al.*, 1999a).

Some of the growth factors, whose important function in the development of BPH has been mentioned before, can be deliberately changed in certain cells by molecular biological methods. Such methods, permitting a programmed change in the DNA, can also be used for the *in vivo* testing of new drugs. Modified urogenital sinus tissue, in this case induced TGFβ1 overexpression, can be implanted into a mouse whose development is then studied (Yang *et al.*, 1996). This model resembles the one developed by Otto *et al.* (1992) in which human BPH tissue was implanted into the skin of a nude mouse whose growth rate was consequently observed. Nettle root extracts have not yet been studied on the basis of this model developed by Thompson (Yang *et al.*, 1996), but the example suggests a direction of future work.

Therapeutic studies

Clinical studies with nettle root extracts have been carried out since the late 1970s. As time went on, the demands made on the studies were more sharply defined (Table 5.1). Consequently, recent studies are placebo-controlled and correspond to the standards set by the GCP and the I-PSS. Some clinical parameters and the subjective conditions of the patients (symptom score) improved during therapy with nettle root extract. A decreasing volume of the prostate has not been ascertained so far. It would be desirable to conduct further studies with a greater number of patients over a longer period. Such studies might help to bring out the differences between synthetic drugs and phytomedicines.

Table 5.1 Recent clinical studies of *Urtica* in the BPH

Kind of study	Number of patients		Remarks
Open study	10		(Goetz, 1989)
	31		(Ziegler, 1982)
	8		(Frick and Aulitzky, 1987)
	33		(Ziegler, 1983)
	10		(Dunzendorfer, 1984)
Open multicentrique study	5,492		(Tosch and Müssiggang, 1983)
	4,051		(Stahl, 1984)
	4,480		(Friesen, 1988)
	253		(Bauer *et al.*, 1988)
Open long term study	89		(Djulepa, 1982)
	30		(Schönfelder *et al.*, 1982)
	30		(Ziegler, 1987)
	26		(Feiber, 1988)
Double blind study	50	Placebo controlled	(Vontobel *et al.*, 1985)
	79		(Dathe and Schmid, 1987)
	40		(Fischer and Wilbert, 1992)
	41	Placebo controlled, GCP, I-PSS*	(Engelmann *et al.*, 1996)

Note
* Good Clinical Practice (GCP), Internationale Prostata-Symptom-Score (I-PSS).

References

Bauer, H. W., Sudhoff, F., and Dressler, S. (1988) Endokrine Parameter während der Behandlung der benignen Prostathyperplasie mit ERU. In Bauer H. W. (ed.) *Benigne Prostatahyperplasie II, J. Klin. Exp. Urol.*, 19, München, 44–49.

Chung, L. W. K., Matsuura, J., Rocco, A. K., Thompson, T. C., Miller, G. J., and Runner, M. N. (1984) Tissue interactions and prostatic growth: a new mouse model for prostatic hyperplasia, *Ann. New York Acad. Sc.*, 438, 394–404.

Cunha, G., Chung, L. W. K., Shannon, J. M., and Reese, B. A. (1980) Stromal-epithelial interactions in sex differentiation, *Biol. Reprod.*, 22, 19–42.

Dathe, G. and Schmid, H. (1987) Phytotherapie der benignen Prostatabyperplasie (BPH), *Urol. Ausg. B*, 27, 223–226.

Daube, G. (1988) Pilotstudie zur Behandlung der benignen Prostatahyperplasie bei Hunden mit Extractum radicis urticae. In Bauer H. W. (ed.) *Benigne Prostatahyperplasie II, J. Klin. Exp. Urol.*, 19, München, 44–49.

Djulepa, J. (1982) Prostata: So läßt sich die Restharnmenge reduzieren, *Ärztl Praxis*, 63, 2199.

Dunzendorfer, U. (1984) Der Nachweis von Reaktionseffekten des Extractum radicis urticae (ERU) im menschlichen Prostatgewebe durch Fluoreszensmikroskopie, Z. *Phytotherapie*, 5, 800–804.

Enderle-Schmitt, U., Gutschank W.-M., and Aumüller, G. (1988) Wachstumskinetik von Zellkulturen aus BPH unter Einfluß von Extractum radicis urticis (ERU). In Bauer, H. W. (ed.) *Benigne Prostatahyperplasie II, J. Klin. Exp. Urol.*, 19, München, 56–61.

Engelmann, U., Boos, G., and Kres, H. (1996) Therapie der benignen Prostatahyperplasie mit Bazoton-Liquidum, *Urol. Ausg. B*, 36, 287–291.

Feiber, H. (1988) Sonographische Verlaufsbeobachtung zum Einfluß der medikamentösen Therapie der benignen Prostatahyperplasie (BPH). In Bauer, H. W. (ed.) *Benigne Prostatahyperplasie II, J. Klin. Exp. Urol.*, 19, München, 75–82.

Fischer, M. and Wilbert, D. (1992) Wirkprüfung eines Phytopharmakons zur Behandlung der benignen Prostatahyperplasie (BPH). In Rutishauser, G. J. (ed.) *Benigne Prostatahyperplasie III, J. Klin. Exp. Urol.*, 22, München, 79–84.

Frick, J. and Aulitzky, W. (1987) Auswertung von Hormon- und Samenmeßgrößen bei Patienten mit Stauungszuständen der Prostata, die mit Extractum radicis urticae behandelt wurden. In Bauer H. W. (ed.) *Benigne Prostatahyperplasie, J. Klin. Exp. Urol.*, 14, München, 48–51.

Friesen, A. (1988) Statistische Analyse einer Multizenter-Langzeitstudie mit ERU. In Bauer, H. W. (ed.) *Benigne Prostatahyperplasie II, J. Klin. Exp. Urol.*, 19, München, 121–130.

Ganßer, D. and Spiteller (1995) Aromatase inhibitors from *Urtica dioica* roots, *Planta Medica*, 61, 138–140.

Geiger, W. N., Haak, C., and Wagner, H. (1996) Receptor binding assay with Urtica dioica agglutinin (UDA) and investigation of the uptake of this iodine 125 labelled lectin in mice. In Wagner, H. (ed.) *2nd International Congress on Phytomedicine*, München, sl–106. (6)

Geiger, W. N., Haak, C., and Wagner H. (1997) Urtica dioica-Agglutinin (UDA): a pharmacologically active principle of *Urtica dioica* phytopreparations. In Franz, G. and Vieweger, U. (ed.) *45th Annual Congress of the Society of Medicinal Plant Research*, Regensburg, I–11. (6)

Goetz, O. (1989) Die Behandlung der benignen Prostatahyperplasie mit Brennesselwurzeln, *Z. Phytotherapie*, 10, 175–178

Hirano, T., Homma, M., and Oka, K. (1994) Effects of stinging nettle root and their steroidal components on the Na$^+$,K$^+$-ATPase of the benign prostatic hyperplasia, *Planta Medica*, 60, 30–33.

Hryb, D. J., Khan, M. S., Romas, N. A., and Rosner W. (1995) The effect of extracts of the roots of stinging nettle (*Urtica dioica*) on the interaction of SHBG with its receptor on human prostatic membrans, *Planta Medica*, 61, 31–32.

Koch, E. (1995) Pharmakologie und Wirkmechanismen von Extrakten aus Sabalfrüchten (Sabal fructus), Brennesselwurzeln (Urticae radix) und Kürbissamen (Curcurbitae peponis semen) bei der Behandlung der benignen Prostatahyperplasie. In Loew, D. and Rietbrock, N. (ed.) *Phytopharmaka in der Forschung und klinischen Anwendung, Darmstadt*, 57–79.

Koch, E. (2001) Extracts from fruits of saw palmetto (*Sabal serrulata*) and roots of stinging nettle (*Urtica dioica*): viable alternatives in the medical treatment of benign prostatic hyperplasia and associated lower urinary tracts symptoms, *Planta Medica*, 67, 489–500.

Koch, E., Jaggy, H., and Chatterjee, S. (1995) Inhibition of human leukocyte elastase by an ethanolic extract from roots of stinging nettle (*Urtica dioica* L.), *Naunyn-Schmiedeberg's Arch Pharmak*, 351, R57. (2)

Konrad, L., Müller, H.-H., Lenz, C., Laubinger, H., Aumüller, G. and Lichius, J. J. (2000) Antiproliferative effects of Stinging nettle root extract on human prostate cancer cells, *Planta Medica*, 66, 44–47.

Kraus, R. (1990) Inhaltsstoffe der Extrakte von *Urtica dioica* und ihre potentielle Wirksamkeit, Thesis, Bayreuth.

Lichius, J. J. and Muth, C. (1997) The inhibiting effects of *Urtica dioica* root-extracts on experimentally induced prostatic hyperplasia in the mouse, *Planta Medica*, 63, 307–310.

Lichius, J. J., Renneberg, H., Blaschek, W., Aumüller, G., and Muth, C. (1999a) The inhibiting effects of components of Stinging nettle roots on induced prostatic hyperplasia in mice, *Planta Medica*, 65, 666–668.

Lichius, J. J., Lenz, C., Lindemann, P., Aumüller, G., and Konrad, L. (1999b) Antiproliferative effect of a polysaccharide fraction of a 20% methanolic extract of Stinging nettle roots upon epithelial cells of the human prostate (LNCaP), *Die Pharmazie*, 54, 768–771.

McNeal, J. E. (1978) Origin and evolution of benign prostatic enlargement, *Invest. Urol.* (Baltimore), 15, 340–345.

Otto, U., Wagner, B., Becker, H., Schröder, S., and Klosterhalfen, H. (1992) Transplantation of human benign hyperplastic prostate tissue into mice: first results of systemic therapy, *Urol. Int.*, 48, 167–170.

Rausch, U., Aumüller, G., Eicheler, W., Gutschank, W., Beyer, G., and Ulshofer, B. (1992) Der Einfluß von Phytopharmaka auf BPH-Gewebe und Explantatkulturen *in vitro*. In Rutishauser, G. J. (ed.) *Benigne Prostatahyperplasie III*, *J. Klin. Exp. Urol.*, 22, München, 116–124.

Rhodes, L., Primka, R. L., Berman, C., Vergult, G., Gabriel, M., Pierre-Mallce, M., and Giblein, B. (1993) Comparison of finasterid (Proscar®), a 5α-reductase inhibitor, and various commercial plant extracts in *in vitro* and *in vivo* 5α-reductase inhibition, *Prostate*, 22, 43–51.

Rosner, W. (1994) New insights into the function of SHBG. In Boos, G. (ed.) *Benigne Prostatahyperplasie*, Frankfurt, 87–95.

Rückle, E. (1950) Brennesselwurzeltee bei beginnender Prostatitis, Hippokrates: Informationen aus der Medizin für naturgemässe Lebens- und Heilweise, *Wissenschaft und Praxis*, 21, 55–56.

Scapanini, U. and Friesen, A. (1992) *Urtica dioica*-Extrakt und Folgesubstanzen im Tierversuch. In Rutishauser G. J. (ed.) *Benigne Prostatahyperplasie III*, *J. Klin. Exp. Urol.*, 22, München, 138–144.

Schmitt, J., Gutschank, M., Heck, H., Enderle-Schmitt, U., and Aumüller, G. (1987) Cell culture of prostatic stromal tissue. In Bauer, H. W. (ed.) *Benigne Prostatahyperplasie*, *J. Klin. Exp. Urol.*, 14, München, 18–22.

Schönfelder, G., Tauber, R., Rattenhuber, U., and Barth, H. (1982) Die Bedeutung von Docosanol und esetractum Radicis Urticae zur konservativen therapie der protatahyperplasie. *J. Klin. Exp. Urol.* 4, 179–182.

Schöttner, M., Ganßer, D., and Spiteller, G. (1997) Lignans from the roots of *Urtica dioica* and their metabolites bind to human sex hormone binding globulin (SHBG), *Planta Medica*, 63, 529–532.

Stahl, H.-P. (1984) Die Therapie prostatischer Nykturie, *Z. Allg. Medizin*, 3, 128–132.

Tosch, U. and Müssiggang H. (1983) Medikamentöse Behandlung der benignen Prostatahyperplasie, *Euromed.*, 6, 1–3.

Vontobel, H. P., Herzog, R., Rutishauser, G., and Kres, H. (1985) Ergebnisse einer Doppelblindstudie über die Wirksamkeit von ERU-Kapseln in der konservativen Behandlung der benignen Prostatahyperplasie, *Urol. Ausg. A*, 24, 49–51.

Wagner, H., Willer, F. and Kreher, B. (1989) Biologisch aktive Verbindungen aus dem Wasserextrakt von *Urtica dioica*, *Planta Medica*, 55, 452–454.

Wagner, H., Willer, F., Samtleben, R., and Boos, G. (1994) Search for the antiprostatic principles of stinging nettle (*Urtica dioica*) roots, *Phytomedicine*, 1, 213–224.

Wagner, H., Geiger W. N., Boos, G., and Samtleben, R. (1995) Studies on the binding of Urtica dioica agglutinin (UDA) and other lectins in an *in vitro* epidermal growth factor receptor test, *Phytomedicine*, 4, 287–290.

Willer, F. (1992) Chemie und Pharmakologie der Polysaccharide und Lektine von *Urtica dioica* L., Thesis, München.

Yang, G., Timme, T. L., Park, S. H., Wu, X., Liu, A., and Thompson, T. C. (1996) A mouse model for human BPH and its use to test the effects of doxazosin *in vivo*. In NIH (ed.) *International Symposium on Biology of Prostate Growth*, Washington, #26, (2).

Ziegler, H. (1982) Zytomorphologische Untersuchungen der Benignen Prostatahyperplasie unter Behandlung mit Extractum radicis urticae, *Fortschr. Med.*, 100, 1832–1834.

Ziegler, H. (1983) Fluoreszenzmikroskopische Untersuchungen von Prostatazellen unter Einwirkung von Extract. radicis urticae (ERU), *Fortschr. Med.*, 101, 2112–2114.

Ziegler, H. (1987) Vorläufige Ergebnisse einer 5-Jahres-Langzeitbehandlung der BPH mit ERU. In Bauer H. W. (ed.) *Benigne Prostatahyperplasie, J. Klin. Exp. Urol.*, 14, München, 23–25.

6   Nutritional evaluation of *Urtica* species

Huriye Wetherilt

Introduction

The consumption of stinging nettle (common nettle) as a vegetable dates back to antique times. In ancient Greece, it was known as 'apokalif' and 'knide' and mentioned in works by Hippocrates (460–377 BCE) and Theophrastus (372–285 BCE). Its use as food is praised and advised in *Materia Medica* by Dioscorides and *Naturalis Historia* by Plinius, both works dating back to the first century CE. In *Materia Medica*, the remedial properties of the plant are also listed with special emphasis on its use for treatment of dog bites, gangrene, rheumatism, ulcers and tumours. In the middle ages, its popularity in folk medicine continued to increase. Recipes for medicinal preparations of the stinging nettle were given in *Physica* by Hildegard von Bingen (1098–1179), and *Signatura Plantarum* by Paracelsus (1493–1541) and advocated in *Dyetary of Helth* by Borde and *Kreütterbuch* by Hieronymus Boch, two well-known herbal medicinal books of the renaissance ages. In the present day, the stinging nettle still retains its importance as a medicinal plant. Its various uses in medicine are given in detail by Madaus (1938) and Keeser (1940) in German school books. Its traditional use in folk medicine is also reviewed by several other workers (Aksoy *et al.*, 1977; Lutomski and Speichert, 1983; Atasu and Cihangir, 1984; Chaurasia, 1987).

Due to its diverse climatic conditions and geographical terrain, Asia Minor enjoys a large variety of plant life, consisting of over 10,000 species. Anatolian people have taken advantage of this variety, using plants as food and medicine since the Paleolithic ages. The stinging nettle is consumed widely as a vegetable by northern and eastern Anatolian people.

The common nettle, which is, as the name suggests, widely distributed around the world, is a green plant with stinging hairs on both faces of its leaves. It refers to several stinging varieties of *Urtica* species, which belong to the family Urticaceae. In the plant kingdom, these varieties come under the division of Spermatophyta, subdivision Angiospermae; class Dicotyledonae; group Apetalae; order Urticales; family Urticaceae; tribus Urticeae; species *Urtica* (Karamanoğlu, 1977). The Urticaceae family comprises some 40 species and 500 varieties of monoic and dioic plants growing in tropical and subtropical regions. The main varieties identified under the *Urtica* species are *Urtica dioica* L., *U. urens* L. or *U. minor* Lam., *U. cannabina* L., *U. pilulifera* L., *U. membranacea* Poir. and *U. kiovensis* Rogoff.

The varieties that are used as spring vegetable are *U. dioica* (dioic, perennial, 30–150 cm long), *U. urens* (monoic, annual, 15–45 cm long) and *U. pilulifera* (monoic, annual or rarely biannual, 30–90 cm long). The part of the plant that is above soil level is called Herba Urticae and that below Radix Urticae. The leaves have single-celled stinging hairs, which come off at the slightest contact and secrete the sap contained in their vacuoles. This sap is a strong irritant due to its histamine and acetylcholine content and thus constitutes an effective defence system for the plant against animals. These varieties are widely distributed in the subtropical areas of Asia, Europe and North America. They are mostly found near inhabited areas, especially in the shade of garden walls or ruins (Baytop, 1963).

The common nettle is used widely as a vegetable in the Black Sea Region of Turkey. It is a very nutritious plant with regard to, especially, high vitamin and mineral contents. Its use as a foodstuff in Anatolia dates back to ancient times and is mentioned in *Materia Medica*, written in the first century CE by the famous physician and botanist Pedanius Dioscorides Anazarbeus, who advises the consumption of nettle leaves, especially after cooking together with sea products, as a remedy for gastric problems (Gunther, 1959). In his *Naturalis Historia*, the Roman statesman Cajus Plinius Secundus, who also lived in the first century, praised the dish made with young nettle leaves collected in spring time (Anon, 1961). Until the twentieth century, nettle leaves were used as a component of soups and salads in Europe (Davies, 1813; Louden, 1836; Merle, 1842). Bryant (1783), in his book on dietetic plants, wrote that the new nettle shoots were collected, boiled and used in salads by many people in early spring.

In present times, the common nettle has lost much of its old popularity in Europe, but owing to its high nutritional value, is still consumed by 'natural food' fans (Nicholson *et al.*, 1969; Mabey, 1975; Grieve, 1976). In Finland, the plant is used commonly as a vegetable (Peura and Koskenniemi, 1985).

The fresh leaves of *Urtica dioica* L.

The leaves of the *Urtica* species have been reported to be excellent sources of some important minerals and vitamins (Booth and Bradford, 1963; Trofimova, 1977; Adamski and Bieganska, 1980) and to have a higher level and better quality of protein when compared with many other green leafy vegetables (Ullrich and Jahn-Deesbach, 1984; Wetherilt, 1989).

The leaves of the stinging nettle contain ample chlorophylls and carotenoids. Aro *et al.* (1986) have found 4.8 mg chlorophyll per gram of dry leaves. However, because the nettle is a photolabile plant, its chlorophyll and carotenoid contents change depending on whether the plant has grown in the sun or shade. Popova *et al.* (1982) have compared the two conditions and found more chlorophyll and carotenoids in plants grown in the shade. Kudritskaya *et al.* (1986) found the carotenoids fraction of *Urtica dioica* to consist of mainly β-carotene, violaxanthin, xanthophyll, zeaxanthin, luteoxanthin and lutein epoxide.

The dry matter content of the fresh leaves grown in Turkey was found to be 23% (Wetherilt, 1992) and 23% and 24% by different workers in Germany (Seybold and

Table 6.1 Amino acid composition of the fresh leaf protein hydrolysate (g/100 g), chemical score and real protein value (%)

Amino acid	Wetherilt (1992)	Ullrich and Jahn-Deesbach (1984)	Huges et al. (1980)
Phenylalanine	5.82	5.62	6.82
Lysine	5.53	6.97	13.88
Threonine	4.61	4.72	5.40
Valine	6.31	5.81	7.21
Methionine	1.76	1.89	0.87
Cysteine	0.85		
Isoleucine	4.78	4.25	4.91
Leucine	8.97	8.50	7.39
Tryptophan	1.28		
Histidine	4.10	1.90	2.92
Aspartic acid	9.07	12.78	10.78
Serine	6.19	6.05	4.61
Glutamic acid	13.30	12.73	13.09
Proline	4.87	4.91	4.83
Glycine	6.25	4.89	6.59
Alanine	6.54	6.07	6.67
Tyrosine	3.87	3.56	4.03
Arginine	5.90	6.48	
Total essential a.a.	39.21	37.76	
Chemical score	56		
Real protein value	3.1		

Mehner, 1948; Franke and Kensbock, 1981). Wetherilt (1992) found the fresh leaves to contain 76.9% water, 1.6% fat, 6.5% protein, 4.1% nitrogen free extract, 5.3% fibre and 5.6% ash. This protein level corresponds to 28% on a dry matter basis.

Hughes *et al.* (1980) analysed the protein contents of the leaves in every month of the year and found the lowest level, on a dry matter basis, in December (20.9%). Protein levels began to increase regularly after this month and reached peak value in April (36.0%), after which it began to decrease again until January. Adamski and Bieganska (1984) have also studied the protein content of leaves and found them to change from 20.7% before fluorescence to 27.4% after fluorescence. The results of the amino acid composition of nettle leaf protein by different workers are given in Table 6.1. With their high protein and lysine content and real protein value, nettle leaves offer a better quality protein than all other green vegetables. When compared with levels in spinach and parsley, nettle leaves contain three times as much ash and fibre and twice as much protein (Anon, 1982).

Wetherilt (1982) found 100 g fresh leaves (as is) to contain 0.015 mg thiamin, 0.23 mg riboflavin, 0.62 mg niacin and 0.068 mg vitamin B6. Analyses also revealed 238 mg vitamin C, 5 mg β-carotene and 14.4 mg α-tocopherol in 100 g of leaves. These are remarkably high values for these antioxidants with vitamin activities and make nettle leaves an excellent natural source for protection against neoplastic diseases, cardiovascular disorders and immune deficiency. In support of this claim, fresh leaves were found to show antitumoural activity in animal studies and strong antimutagenic activity in the Ames test (Wetherilt, 1988; Karakaya and El, 2000).

Nutrient analyses showed the leaves to be rich in minerals as well, especially with respect to the nutritionally important ones such as iron, calcium and potassium: iron (13 mg/100 g); zinc (0.9 mg/100 g); copper (0.52 mg/100 g); calcium (853 mg/100 g); phosphorus (75 mg/100 g); magnesium (96 mg/100 g); manganese (3 mg/100 mg); sodium (16 mg/100 g); potassium (532 mg/100 g); and selenium (2.7 μg/100 g) (Wetherilt, 1992). The high potassium to sodium ratio (33.2) is another indicator of the protective powers of the *U. dioica* foliage against cardiovascular and neoplastic diseases.

The dried flowers of *Urtica dioica* L.

Samples of dried flowers purchased from herbalists in Turkey were analysed for nutrient contents by Wetherilt (1992). They were found to contain, on average and as is, 11% water, 10.8% fat, 17.2% protein, 31.3% nitrogen free extract, 14.8% fibre and 14.9% ash. The dried flowers were found to be rich in α-tocopherol (16.9 mg/100 g); riboflavin (0.76 mg/100 g); iron (43 mg/100 g); zinc (2.6 mg/100 g); calcium (3.0 g/100 g); phosphorous (400 mg/100 g) and potassium (1.5 g/100 g). However, the analyses also indicated that as a result of the drying and storage process, total loss of vitamin C activity and a substantial loss of β-carotene activity (1.9 mg/100 g) had been incurred.

The amino acid compositions of the dried flowers are given in Table 6.2. In comparison to the composition of the fresh leaves, the dried flower hydrolysate had lower levels of lysine, sulphur amino acids as well as essential amino acids. Accordingly, the chemical score for the dried flowers was also lower.

The seeds of *Urtica pilulifera* L.

A mixture of crushed *U. pilulifera* seeds and honey is used widely in Turkey as a herbal remedy for leukemia; cervical, breast and prostate cancers; and tumours of the mouth, lung and gastrointestinal systems. Nutritional analyses were conducted by Wetherilt (1992) on a mixture of seed samples purchased from four different herbalists in the Istanbul Spice Market. The seeds were found to contain 8.5% water, 25% fat, 21.9% protein, 26.4% nitrogen free extract, 11.4% fibre and 6.8% ash. With a high chemical score and protein value, as calculated from the amino acid composition (Table 6.2), the seeds appear to be a relatively good source of plant protein. However, because of their claimed antitumoural activities, the seeds are usually sold at high prices.

In Table 6.3, the fatty acid compositions of the seed oils from *U. pilulifera* (Wetherilt, 1992) and *U. dioica* (Lotti *et al.*, 1985) are compared. *U. pilulifera* oil has higher levels of saturated (mainly palmitic acid) and monounsaturated (mainly oleic acid) acids, whereas in *U. dioica* oil, the level of the polyunsaturated fatty acid, linoleic acid, is considerably higher.

Table 6.2 Amino acid composition (g/100 g), chemical score and real protein value (%) of the protein hydrolysates from and *U. pilulifera* seeds and dried *U. dioica* flowers

	U. dioica *flowers*	U. pilulifera *seeds*
Phenylalanine	5.51	5.34
Lysine	4.67	5.82
Threonine	4.53	4.78
Valine	5.80	6.42
Methionine	1.39	1.62
Cysteine	0.35	2.21
Isoleucine	4.10	4.56
Leucine	7.18	6.56
Tryptophan	1.91	0.97
Histidine	4.52	4.30
Aspartic acid	13.15	9.20
Serine	6.87	6.75
Glutamic acid	12.32	14.73
Proline	4.21	4.78
Glycine	5.41	6.19
Alanine	6.27	3.15
Tyrosine	2.80	3.42
Arginine	6.13	9.20
Total essential a.a.	35.44	38.28
Chemical score	42	87
Real protein value	6.3	15.5

The seeds of *U. pilulifera* were also found to be fair sources of the following vitamins and minerals: vitamin C (6 mg/100 g); vitamin E (20.2 mg/100 g); thiamin (0.13 mg/100 g); riboflavin 0.22 mg/100 g); niacin (1.79 mg/100 g); vitamin B6 (0.15 mg/100 g); iron (33 mg/100 g); zinc (4.3 mg/100 g); copper (0.90 mg/100 g); calcium (2.17 g/100 g); phosphorous (0.64 g/100 g); magnesium (0.35 g/100 g); manganese (6 mg/100 g); sodium (24 mg/100 g); potassium (0.66 g/100 g); and selenium (0.078 mg/100 g).

Although the seeds are not as rich in micronutrients as the leaves of the plant, their use in folk medicine for treatment of certain cancers makes them a sought after and expensive commodity. On the other hand, the leaves, especially when fresh, are an exceptionally good source of important vitamins and minerals. Moreover, they are available in abundance in many areas of the world for collection and consumption. Particularly in spring, when vitamin sources are scarce, consumption of fresh leaves can serve as a valuable food source for some very important vitamins. However, because the leaves are also known to contain oxalates, which bind the minerals and therefore reduce their absorbance in the intestinal wall, it is not possible to make an estimate of the availability of these minerals to the organism. On the other hand, if the nutrients can be extracted free of oxalates from the nettles, their processing into supplementary drugs could yield nutritional and economic benefits.

Table 6.3 Comparison of fatty acid composition of oils from *U. pilulifera* and *U. dioica* seeds

Fatty acids	U. pilulifera *seed* (*Wetherilt, 1992*)	U. dioica *seed* (*Lotti* et al., *1985*)
propionic acid C 3-0	2.39	—
butyric acid C 4-0	0.13	—
caproic acid C 6-0	0.21	—
caprylic acid C 8-0	0.10	—
capric acid C 10-0	0.09	—
myristic acid C 14-0	0.08	0.05
pentadecanoic acid C 15-0	trace	0.01
pentadecenoic acid C 15-1	0.11	0.02
palmitic acid C 16-0	8.01	3.25
palmitoleic acid C 16-1	0.85	0.05
heptadecanoic acid C 17-0	0.14	0.02
heptadecenoic acid C 17-1	0.17	0.05
stearic acid C 18-0	3.11	0.68
oleic acid C 18-1	23.20	11.20
linoleic acid C 18-2	59.34	81.46
linolenic acid C 18-3	0.82	1.38
arachidic acid C 20-0	0.08	0.21
eicosenoic acid C 20-1	trace	0.20
eicosadienoic C 20-2	—	0.03
eicosatrienoic acid C 20-3	—	0.14
behenic acid C 22-0	1.17	1.25
Saturated acids	15.51	5.47
Monounsaturates	24.33	11.52
Polyunsaturates	60.16	83.01

Recipes using fresh nettle leaves

In view of the exceptional nutritional properties of nettle leaves presented earlier, three recipes are given here with nettles as an ingredient. These are traditional dishes from the rural parts of the Aegean Region of Turkey and are prepared mostly in spring when the nettle leaves are young. Please remember to wear gloves when handling nettles prior to cooking.

Dönderme

Ingredients	Measure	Amount (in g)
nettle leaves		400
egg	1 medium size	50
onion	1 large size	100
flour	4 tablespoons	18
olive oil	3 tablespoons	30
black pepper	1/2 teaspoon	1
salt	1/2 teaspoon	3

Instructions Strip the nettle leaves from their stems and wash and drain well; if young and crisp, the shoots can also be used. Chop the leaves and young shoots coarsely and combine in a large bowl with the finely chopped onion, two tablespoons flour, two tablespoons olive oil, one egg, salt and black pepper. Mix all the ingredients thoroughly.

Put 1/2 tablespoon olive oil in a non-stick pan. After heating gently, sprinkle with flour. Spread the mixture you have prepared onto the pan and push down gently with a spoon to give shape. Cook over low heat while turning the pan slowly to provide even cooking. When the bottom is browned evenly, turn over onto a tray. Again oil the pan with 1/2 tablespoon olive oil and sprinkle with flour. Cook the other side in the same manner until brown. Take the cooked dish on a large plate, cut into squares. Put yogurt and black pepper on top and serve hot.

Spinach and nettles cooked in olive oil

Ingredients	Measure	Amount (in g)
nettle leaves		200
leeks		300
spinach		500
olive oil	1/3 cup	80
black pepper	1/2 teaspoon	1
salt	1/2 teaspoon	6

Instructions Wash nettle leaves, spinach and leeks well and drain. Chop spinach and nettles coarsely and leeks in about 0.5 cm thick slices. Put the olive oil in a pan and add the leeks, spinach, nettles and salt. Cover and heat on high heat until the lid of the pan is too hot to touch. Then bring the heat to the lowest setting and simmer gently without stirring for 20 minutes. Stir gently and cook for another 5 minutes. Serve hot or cold according to taste with yogurt and black or red pepper.

Spinach and nettle pie

Ingredients	Measure	Amount (in g)
spinach		400
nettle leaves		200
milk	1 cup	250
eggs	2 medium size	100
onion	1 medium size	60
flour	3 tablespoons	18
olive oil	4 tablespoons	60
butter or margarine	1/4 cup	60
black pepper	1/2 teaspoon	1
salt	1 teaspoon	6
fyllo pastry (yufka)	4 sheets	600

Instructions For filling (Mixture A): Wash nettle leaves and spinach well, drain and shred or chop coarsely. Sauté the chopped onion in olive oil on medium heat until transparent but not brown. Add nettles and spinach, stir and cook over low heat for 10 minutes. Remove from heat, season with salt and pepper and cool. Stir in one beaten egg and set aside. Mixture B: Beat and mix melted butter, two tablespoons oil, milk and one egg thoroughly.

Butter a large baking pan (round or square) and lay one sheet of fyllo pastry onto it. Trim the sides of pastry to fit into the baking pan. Brush with mixture B. Spread 1/3 of mixture A on top. Put another layer of fyllo pastry on top and apply mixture B and A in the same manner until the final fyllo layer is laid. Fold overhanging edges over top layer, brush liberally with mixture B. Bake in a moderate oven for 30 minutes or until golden brown. Cut into squares to serve hot or sprinkle with one tablespoon water and let stand covered until warm or cold.

Acknowledgement

The author would like to thank Mrs Günsu Çirpanli Albaş for providing the traditional recipes.

References

Adamski, R. and Bieganska, J. (1980) Studies of chemical substances present in *Urtica dioica* L. leaves. Part 1. Trace elements. *Herba Pol.* 26(3), 177–180.

Adamski, R. and Bieganska, J. (1984) Studies on substances present in *Urtica dioica* L. leaves. Analysis for protein, amino acids and nitrogen containing nonprotein substances. *Herba Pol.* 30(1), 17–26.

Aksoy, C., Yücecan, S., Çiftçi, N., Tayfur, M., Akgün, B., and Taşcı, N. (1988) Kanser hastalığında tedavi amacıyla kullanılan yöresel bitkiler. *Beslenme ve Diyet Dergisi* 17, 111–120.

Anon (1961) *Pliny Natural History Book* XXI, Vol. VI, Heinemann Publications, London, p. 229.

Anon (1982) *Food Composition Tables for the Near East.* Food and Agriculture Organization of the United Nations, Rome.

Aro, E. M., Rintamaki, E., Karhonen, P. and Maenpaa, P. (1986) Relationship between chloroplast structure and O_2 evolution rate of leaf disks in plants from different biotopes in South Finland. *Plant, Cell Environ.* 9, 87–94.

Atasu, E. and Cihangir, V. (1984) *Urtica* L. türlerinin kimyasal içeriği ve tedavide kullanimi. *FABAD* 9(2), 73–81.

Baytop, T. (1963) Tibbi ve Zehirli Bitkiler. İstanbul Üniversitesi Eczacilik Fakültesi Farmakognozi Bölümü, İsmail Akgün Matbaasi, 119.

Booth, V. H. and Bradford, M. P. (1963) Tocopherol contents of vegetables and fruits. *Br. J. Nutr.* 17, 575–581.

Bryant, C. (1783) *Flora diatetica* or *History of Esculant Plants.* White Publications, London, p. 133.

Chaurasia, N. (1987) Phytochemische, Untersuchungen einiger *Urtica*-Arten unter besonderer Berücksichtigung der Inhaltsstoffe von Radix und Flores *Urticae dioica* L. Inaugural-Dissertation zur Erlangung der Doktorwürde. Fachbereichs Pharmazie und Lebensmittelchemie der Philipps-Universitat, Marburg/Lahn.

Davies, H. (1813) *Welsh Botanology*. W. Marchant Publishers, London, p. 171.

Franke, W. and Kensbock, A. (1981) Vitamin C Gehalte von Heimischen Wilgemusearten und Wildsalatarten. *Ernaehr. Umsch* 28(6), 187–191.

Grieve, M. (1976) *A Modern Herbal*. Penguin Books, Harmondsworth Publications, p. 574.

Hughes, R. E., Ellery, P., Harry, T., Jenkins, V. and Jones, E. (1980) The dietary potential of the common nettle. *J. Sci. Food Agric.* 31, 1279–1286.

Gunther, R. T. (1959) *The Greek Herbal of Dioscarides*. Hafner Publications, New York, p. 491.

Karakaya, S. and El, S. N. (2000) Determination of antimutagenic effects of some foods and drinks *in vitro* binding capacities of some dietary fibers to mutagens. *J. Nutr. And Diet* 29(2), 4–13.

Karamanoğlu, K. (1977) Farmasotik Botanik. Ankara Üniversitesi Eczacilik Fakültesi Yayınları No. 44, Ankara Üniversitesi Basımevi, Ankara.

Keeser, E. (1940) Zür Wirkungsweise der brennessel. *Deut. Med. Wochenschrift*, 66, 849.

Kudritskaya, S. E., Fishman, G. M., Zagorodskaya, L. M. and Chicovani, D. M. (1986) Carotenoids of *Urtica dioica*. *Khim. Prir. Soedin.* 5, 640–641.

Lotti, G., Paradossi, C. and Marchini, F. (1985) *Riv. Soc. Ital. Sci. Aliment.* 14, 263–270.

Loudon, J. C. (1836) *An Encyclopedia of Plants*. Longman Publications, London.

Lutomski, J. and Speichert, H. (1983) Stinging nettle in medical science and nutrition. *Pharm. Unserer Zeit.* 12(6), 181–186.

Mabey, F. (1975) *Food for Free*. Fontana/Collins Publications, Glasgow, p. 100.

Madaus, G. (1938) Lehrbuch der Biologischen Heilmittel Abt. I.: Heilpflanzen, Bd. I and II., Thieme Verlag, Leipzig.

Merle, G. (1842) *The Domestic Dictionary and Housekeeper's Manual*. William Strange Publications, London, p. 160.

Nicholson, B. E., Harrison, S. G., Masefield, G. B. and Wallis, M. (1969) *The Oxford Book of Food Plants*. Oxford University Press, p. 190.

Peura, P. and Koskenniemi, J. (1985) Simultaneous determination of nitrate and nitrite by ion chromatography in leaves of *Urtica dioica* L. (Urticaceae) in Finland. *Acta Pharm. Fenn.* 94, 67–70.

Popova, I. A., Maslova, T. G., Popova, O. F., Miroslavov, E. A. and Tsarkova, V. A. (1982) Characteristics of the photosynthetic apparatus of stinging nettle growing under various light conditions. *Fiziol. Rast.* 29(6), 1102–1108.

Seybold, A. and Mehner, H. (1948) Über den Gehalt an Vitamin C in Pflanzen. Sitzgsber. Heidelberg Akad. Wiss., Math-Naturwiss. Klasse 10. Abhdlg., Springer Verlag, Heidelberg.

Trofimova, E. P. (1977) Some wild food plants of Tadzhikistan as sources of vitamins. *Izv. Akad. Nauk Tadzh SSR, Otd. Biol Nauk* 1, 43–48.

Ullrich, I. and Jahn-Deesbach, W. (1984) Proteinggehalt und Proteinzusammen-setzung verschiedener Unkrautarten. *Angew. Bot.* 58(3–4), 255–66.

Wetherilt, H. (1989) Assessment of the nutritional properties and antitumoural activity of the common nettle grown in Turkey. PhD thesis. Hacettepe University, Ankara.

Wetherilt, H. (1992) Evaluation of *Urtica* species as potential sources of important nutrients. In G. Charalambous (ed.) *Food Science and Human Nutrition*. Elsevier Science Publishers, pp. 15–25.

7 *Urtica* products

Gulsel Kavalali and Colin Randall

Plant sources

Mostly *Urtica dioica* L. (common nettle) but occasionally also *Urtica urens* L. (small nettle) Family, Urticaceae.

Other names

Urtica dioica L.: nettle, stinging nettle, greater nettle, nettle wort, urtica (Engl.) Brennessel kraut, Haarnessel kraut, Hanfnessel kraut (Ger.) Herbe d'ortie, Grande ortie (Fr.), Ortica (Ital.), Ortiga (Sp.), Zwyczajna (Polish) and Isirgan (Turkish).
   Urtica urens L: small nettle and dwarf nettle..

Parts used

Urtica radix (nettle root), Urtica herba (nettle herb), Urtica fructus (semen) 'nettle fruit (seeds)' and the dried aeriel parts or leaves of *U. dioica*. The sting of the stinging nettle (urtication) has also been used in herbal medicine.

Urtica radix (nettle root)
A number of studies have produced evidence of symptomatic relief of enlarged prostate, frequent urination and weak urinary flow. It is listed as a diuretic in the British Herbal Pharmacopoeia, 1996.

Urtica herba (nettle herb)
It is considered to be a diuretic and also a remedy for rheumatism and gout. The extracts are claimed to increase uric acid excretion. The stinging leaves are applied externally to treat neuralgic and rheumatic pains. They are also applied externally as a remedy for greasy hair and dandruff.
   There is only very limited use of the nettle in the UK by herbalists at the present time. Simon Mills, Director, Department Herbal Medicine, Department Complementary Medicine, Exeter University states: 'It is used occasionally for allergic skin conditions, chronic inflammatory conditions (e.g. arthritis), and as a low level food supplement with high mineral content and detoxicant. He uses nettle juice, fresh pressed herb.' (Mills, personal communication, 2000).
   It is used for allergic rhinitis in the USA and its use is supported by a small clinical randomised placebo controlled trial (Mittman, 1990).

Urtica fructus (semen) 'nettle fruit (seeds)'

They are applied externally as a remedy for skin complaints and rheumatism, but there is little scientific evidence to support this use (Bissett, 1994). They are given internally as a tonic and are said to increase the activity of the liver.

Side-effects

Taking the aeriel part of the plant may cause oedema, skin reactions and gastric irritation. The root sometimes causes mild stomach and intestinal problems.

Drug interaction

No drug interaction has been reported. However, it should not be taken during pregnancy and breast-feeding, and should be avoided in combination with 'Digitalis'. In view of its diuretic action it should be avoided by those taking anti-hypertensive medications which it might augment. It has been variously claimed to increase and decrease blood sugar levels and should be avoided in diabetes (Newall, 1990).

How to prepare

Urtica radix (nettle root)

To make tea: Mix one teaspoonful of coarsely powdered root with a cupful of cold water and boil for one minute, cover and steep for ten minutes and then strain.

Urtica herba (nettle herb)

To make tea: Mix two spoonfuls of finely cut herb with a cupful of cold water, bring briefly to the boil, steep for ten minutes and then strain.

For external use: Traditionally the underside of fresh leaves are applied directly to the painful joints or muscles to cause stinging (urtication) and subsequent pain relief (Randall *et al.*, 2000; Randall, 2001). Alternatively alcoholic extracts made at the same dosage of liquid extract (1 : 1 in 25% alcohol) 3–4 ml, three times daily or tincture (1 : 5 in 45% alcohol) 2–6 ml, three times daily are applied.

Urtica fructus (semen) 'nettle fruit (seeds)'

To make tea: Add water to 2–4 g of the dried and crushed seeds, then boil and stand for ten minutes and strain.

Dose

Tea:	Three cupfuls are taken daily
Dried herb:	2–4 g or by infusion three times daily
Liquid extract:	(1 : 1 in 25% alcohol) 3–4 ml three times daily
Tincture:	(1 : 5 in 45% alcohol) 2–6 ml three times daily
Herbal preparation:	The drug is available in tea bags (1.0–1.8 g)

Table 7.1 Products containing *Urtica* species

Products	Preparation	Main manufacturers	Major therapeutic uses
Arthrodynat	Herba urtica and other plants	Ziethen, Ger.	Arthritis
Bazoton	Radix urtica	Kanolt, Ger.	Prostatic disorders
Befelka – Tinktur	Herba urtica and other plants	Befelka, Ger.	Skin disorders, metabolic disorders, circulatory disorders
Cholaflux	Folium urtica and other plants	Nattermann Tee Arznei, Ger.	Herbal tea
Colchicum – Strath	Herba urtica and other plants	Strath – Labor, Ger.	Joint disorders
Combudoron	Herba urtica and other plants	Weleda, Ger.	Skin disorders
Combudoron	*Urtica urens* and other plants	Weleda, UK	Homoeopathic preparation
Crinocedin	Stinging nettle and other plants	Wolfer, Ger.	Hair tonic
Dr Grandel Brennessel 'Vital Tonicum'	Folium urtica cum semine	Grandel – Synpharma, Ger.	Tonic
Fragador	*Urtica dioica* leaves and other plants	Weleda, UK	Stress
Heparchofid S	Herb. et sem.	Fides, Ger.	Digestive disorders, liver and gall disorders
	Urtica and other plants		
Kleer	Stinging nettles and other plants	Modern Health Products, UK	Skin disorders
Liruptin	Herba urtica and other plants	Fides, Ger	Urinary-tract disorders
Nephropur	Herba urtica and other plants	Repha, Ger.	Urinary-tract disorders
Nosenil	*Urtica* and other plants	Farbo, Ital.	Not mentioned
Prostaforton N	Stinging nettle root	Plantorgan, Ger.	Prostatic disorders
Prostagalen N	Radix *Urtica dioica* and other plants	Galenika, Ger.	Prostatic disorders

(*Continued*)

Table 7.1 (Continued)

Products	Preparation	Main manufacturers	Major therapeutic uses
Prostagutt	Urtica dioica	Schwabe, Ger.	Prostatic disorders
Prostaherb cum 'Belladonna N'	Radix urtica and other plants	Redel, Ger.	Prostatic disorders
Prostatin N	Radix urtica and other plants	Kanoldt, Ger.	Urinary-tract disorders
Prostatin N-Liquidum	Radix urtica and other plants	Kanoldt, Ger.	Urinary-tract disorders
Rheuma – Tea	Herba urtica and other plants	Stada, Ger.	Rheumatic disorders
Rheumex	Herba urtica and other plants	Labopharma, Ger.	Rheumatic disorders
Salus Kurbis – Tonikum	Herba urtica cum radix and other plants	Salushaus, Ger.	Prostatic disorders
Salus Rheuma u. Stuffwechsel Funktionstee Nr.12	Folium urtica and other plants	Salushaus, Ger.	Rheumatic disorders
Salus Rheuma-Tee Krautertee Nr.12	Folium urtica and other plants	Salushaus, Ger.	Rheumatic disorders
Secerna	Herba et sem. urtica and other plants	Fides, Ger.	Tonic
Simic	Urtica root	Zyma, Switz.	Benign prostatic hypertrophy
Stoffwechseldragees	Herba urtica and other plants	Molitor, Ger.	Digestive system disorders
Tisane antirhumatismale 'H'	Herba urtica and other plants	Hanseler, Switz.	Musculoskeletal and joint disorders
Tisane d'allairement 'H'	Urticae herba and other plants	Hanseler, Switz.	Low appetite
Uvirgan	Radix urtica and other plant	Kanoldt, Ger.	Urinary-tract disorders
Vollmers präparierter grüner	Herba urtica and other plants	Salush-aus, Ger.	Urinary-tract disorders

Source: Martindale, 1993.

Table 7.2a Products containing *Urtica* species

Products	Preparation	Main manufacturers	Major therapeutic uses
Urtica plus	Urtica (dried aerial parts of leaves of *U. dioica*)	Euro pharm, Aust.	Urinary-tract disorders
Urtica plus N	Urtica (dried aerial parts of leaves of *U. dioica*)	Hayer, Ger.	Benign prostatic hyperplasia
Urtica prostat Uno	Urtica (dried aerial parts of leaves of *U. dioica*)	Azupharma, Ger.	Benign prostatic hyperplasia
Urticur	Urtica (dried aerial parts of leaves of *U. dioica*)	Biocur, Ger.	Benign prostatic hyperplasia
Urtipret	Urtica (dried aerial parts of leaves of *U. dioica*)	Bionorica, Ger.	Stage I/II prostatic adenoma

Source: Martindale, 1999.

Table 7.2b Markets for products containing *Urtica* species

Countries	Products
Austria	Urtica plus
Germany	Arthrodynat N
	Bazoton
	Cirkuprostan
	Cletan
	Dr Grandel Brennessel Vital Tonikum
	Kneipp Brennessel-Pflanzensaft
	Kneipp Pflanzendragees Brennessel
	Logomed Prostata-Kapseln
	Prostaforton
	Prostagalen
	Prostaherb N
	Prostaneurin
	Prostawern
	Reumaless
	Rheuma-Hek
	Uriginex urtica
	Uro-Pos
	Urtica Plus N
	Urticaprostat Uno
	Urticur
	Urtipret
	Utk Uno
Italy	Venustas Shampo per Capellicon Forfora e/o Grassi
Switzerland	Simic
USA	Nettle capsules
	Nettle Liquid extract

Source: Martindale, 1999; Pitter, 2001.

Table 7.2c Markets for multi-ingredient *Urtica* products

Countries	Products
Austria	Anaemodoron
	Apotheker Bauer's Harntreibender Tee
	Apotheker Bauer's Nieren-und Blasentee
	Berggeist
	Bio-Garten Tee zur Erhohung der Harnmenge
	Bogumil – tassenfertiger milder Abfurtee
	Brennessel tonikum
	Brostalin
	Ehrmann's Ent – schlackungstee
	Fruhjahrs – Elixier ohne Alkohol
	Harntreibender Tee
	Krauterdoktor Entwasserungs – Elixier
	Krauterhaus Mag Kottas Blutreinigungstee
	Krauterhaus Mag Kottas Entschlackungstee
	Krauterhaus Mag Kottas Entwasserungstee
	Krauterhaus Mag Kottas Fruhjahrs-und Herbstkurtee
	Krautertee Nr. 19
	Krautertee Nr. 8
	Mag Doskar's Nieren und Blasentonikum
	Mag Kottas Entschlackungstee
	Menodoron
	Mentopin
	Prostagutt
	Prostatonin
	Sidroga Stoffwechseltee
	Spicies Carvi comp.
	St Radegunder Entwasserungs – Elixier
	St Radegunder Entwasserungstee
	Synpharma Instant – Blasen – und-Nierentee
Australia	Cough relief
	Infant Tonic
	Irontona
	Respatona
	Respatone Plus with Echinacea
	Vitatona
France	Lithiabyl
Germany	Befelka – Tinktur
	C 34 – Strath
	Cholaflux
	Colchicum – Strath
	Combudoron
	Crinocedin
	Dr Klinger's Bergischer Krautertee
	Abfuhr-und Verdauungstee
	Heparchofid S
	Liruptin
	Prostagutt Forte
	Prostagutt Tropfen

(Continued)

Table 7.2c (Continued)

Countries	Products
	Prostatin F
	Prostatin N Liquidum
	Prostatin N
	Rheumex
	Salus Kurbis-Tonikum Compositum
	Salus Rheuma – Tee Krautertee Nr. 12
	Secerna
	Stoffwechseldragees
	Uvirgan N
	Vollmers praparierter gruner
Italy	Herbavit
	Nosenil
	Omadine
	Shamday Antiforfora
Switzerland	Prostagutt-F
	Prostatonin
	Tisane d'allaitement 'H'
	Tisane diuretique
UK	Fragador
	Kleer
	Menodoron

Source: Martindale, 1999.

Table 7.3a Markets for *Urtica* products

Countries	Products
Austria	Urtica Plus
Brazilia	Bazoton
Germany	Anthrodynat N
	Bazoton
	Cirkuprostan
	Cletan
	Dr Grandel Brennesel Vital Tonikum
	Hostid
	Kneipp Brennessel Kraut Pflanzensaft kneippianum
	Kneipp Pflanzen-Drages Brennessel
	Logomed Prostate-Kapsein
	Prostaforton
	Prostagalen
	Prostaherb N
	Prostaneurin
	Prostata

(Continued)

Table 7.3a (Continued)

Countries	Products
	Prostawern
	Reumales
	Rheuma-Hartkapseln
	Rheuma-Hek
	Serless
	Urginex Urtica
	Uro-Pos
	Urtica Plus N
	Urticaprostat uno
	Urticur
	Urupret
	Urtivit
	Utk
Italy	Venustas Shampoo per Capelli con Forfora e/o Grassi
Switzerland	Simic
	Valverde Prostatae capsules
Thailand	Urtipret

Source: Martindale, 2002.

Table 7.3b Markets for multi-ingredient *Urtica* products

Countries	Products
Austria	Anaemodoron
	Apotheker Bauer's Harntreibender Tee
	Apotheker Bauer's Nieren-und Blasentee
	Berggeist
	Bio-Garten Tee zur Erhohung der Harnmenge
	Bogumil-tassenfertiger milder Abfurtee
	Brennessel tonikum
	Brostalin
	Ehrmann's Entschlackungstee
	Florissamol
	Fruhjahrs-Elixier ohne Alkohol
	Harntreibender Tee
	Krauterdoktor Entwasserungs-Elixier
	Krauterhaus Mag Kottas Blutreinigungstee
	Krauterhaus Mag Kottas Entschlackungstee
	Krauterhaus Mag Kottas Entwasserungstee
	Krauterhaus Mag Kottas Fruhiahrs-und Herbstkurtee
	Krautertee Nr.19
	Krautertee Nr.8
	Mag Doskar's Nieren und Blasentonikum
	Mag Kottas Entschlackungstee
	Menodoron

(Continued)

Table 7.3b (Continued)

Countries	Products
	Mentopin
	Naturland Rheuma-Tee
	Prostagutt
	Prostatonin
	Sidroga Stoffwechseltee
	Species Carvi comp.
	St Radegunder Entwasserungs-Elixier
	St Radegunder Entwasserungstee
	Synpharma Instant-Blasen und Nierentee
Australia	Cough Relief
	Extralife Flow-Care
	Infant Tonic
	Irontona
	Respatona Decongestant Formula
	Respatona Plus Bronchial Cough Relief
	Urapro
	Vitatona
Brazilia	Lactifero
	Prostem Plus
France	Lithiabyl
	Salucur
Germany	Befalka-Tinktur
	Combudoron
	Presselin Nieren-Blasen K 3
	Prostagutt Forte
	Prostatin F
	Salus Rheuma-Tee Krautertee Nr.12
	Uvirgan N
	Vollmers praparierter gruner Winar
Italia	Herbavit
	Nosenil
	Omadine
	Shamday Antiforfora
Switzerland	Prostagutt F
	Prostatonin
	Tisane Diuretiqué
UK	Fragador
	Kleer
	Menodoron
	Savlon Natural First Aid for Burns

Source: Martindale, 2002.

Table 7.4 Products containing *Urtica* species in the German market

Products	Preparation	Main manufacturers
Arthrodynat N Tropfen	Mono Wirkstoffkurzbez.: Brennessel	Ziethen
Azuprostat® Urtica	Mono Wirkstoffkurzbez.: Brennessel	Azupharma
Bazoton® N; -uno	Mono Wirkstoffkurzbez.: Brennessel	Abbott
Combudoron®	Combination	Weleda
Desensib®	Combination	Wiedemann
Florabio naturreiner Heilpflanzensaft Brennessel	Mono Wirkstoffkurzbez.: Brennessel	Florabio
Hevertnier Complex	Combination	Hevert
Hewesabal comp.	Combination	Hevert
Heweurat Harnsäuretropfen	Combination	Hevert
Hostid®	Mono Wirkstoffkurzbez.: Brennessel	Fournier Pharma
Hox alpha	Mono Wirkstoffkurzbez.: Brennessel	Strathmann
Lymphtropfen S.	Combination	Cosmochema
Mulimen	Combination	FidesLine
Nieren-Elixier ST	Combination	Cosmochema
Pasisana	Combination	Riemser
Phöno Arnica comp. Salbe	Combination	Phönix
Presselin Nieren-Blasen-Tabletten K 3	Combination	Presselin
Pro-Sabona	Mono Wirkstoffkurzbez.: Brennessel	Sabona
Prostaforton®	Mono Wirkstoffkurzbez.: Brennessel	Biocur
Prostagalen®	Mono Wirkstoffkurzbez.: Brennessel	Galenika Hetterich
Prostagutt® forte	Combination	Schwabe
Prostaherb® N	Mono Wirkstoffkurzbez.: Brennessel	Cesra
Prostaneurin®	Mono Wirkstoffkurzbez.: Brennessel	Sanofi-Synthelabo

Product	Type	Manufacturer
Prostata STADA®	Mono Wirkstoffkurzbez.: Brennessel	Stada
Prostatin F	Combination	Abbott
Prostawern Urtica Liquidum	Mono Wirkstoffkurzbez.: Brennessel	Pharma Wernigerode
Rheuma – Hek®	Mono Wirkstoffkurzbez.: Brennessel	Strathmann
SE Brennnessel	Mono Wirkstoffkurzbez.: Brennessel	Spitzner
Serless®	Mono Wirkstoffkurzbez.: Brennessel	Trum
Solidagosan N	Combination	Hanosa
Toxicerna®	Combination	FidesLine
Uro-POS®	Mono Wirkstoffkurzbez.: Brennessel	Ursapharm
Urtica APS®	Mono Wirkstoffkurzbez.: Brennessel	Aps
Urtica-Hevert Rheumatropfen	Mono Wirkstoffkurzbez.: Brennessel	Hevert
Urticalcin	Combination	Bioforce
Urtica N	Mono Wirkstoffkurzbez.: Brennessel	Hoyer-Madaus
Urticaprostat® uno	Mono Wirkstoffkurzbez.: Brennessel	Azupharma
Urtipret®	Mono Wirkstoffkurzbez.: Brennessel	Bionorica
Urtivit®	Mono Wirkstoffkurzbez.: Brennessel	Bional
Utk®	Mono Wirkstoffkurzbez.: Brennessel	TAD Pharma
utk uno® 460 mg Filmtabletten B	Mono Wirkstoffkurzbez.: Brennessel	TAD Pharma
Uvirgan® N	Combination	Abbott
Vollmers präparierter grüner Hafertee N	Combination	Salushaus
Winar®	Mono Wirkstoffkurzbez.: Brennessel	Apogepha

Source: Rote list, 2002.

Phytomedicines

Urtica semen: Especially alcoholic extract 'Vital – Tonikum[R] Grandel'
Urtica herba: Herbal mixture called 'Blood purifying tea(s)'
Urtica radix: Extractum Radicis urticae-ERU 'Bazoton[R] – Kapseln'

Here in this chapter, an overview on the properties of the *Urtica* products were reviewed and summarized from Bisset (1994), Newall *et al.* (1996), Weiss and Fintelmann (2000), and WHO monograph on selected medicinal plants (2002).

Pharmacopoeia

Urtica preparations and official formularies are listed in some pharmacopoeias. *Urtica dioica* and *Urtica urens* L. or mixtures of these drugs are approved by some pharmacopoeias.

Nettle leaf in DAB 10 (German pharmacopoeia), nettle herb in Pharmacopoeia Helvatica, DAC 1986 (German Drug Formulary) and BHP, 1983 and 1/1990 (British Herbal Pharmacopoeia).
Extract of nettle root is listed in German Commission E Monograph, 1986; Extract of nettle herb is listed in German Commission E Monograph, 1987.
ESCOP, 1996 (revised in 2001 – European Scientific Cooperative on Phytotheraphy) has monographs on both Urtica folium/herba and Urtica radix.
Martindale 28 (1982) The Extra Pharmacopoeia mentions that *U. dioica* plant contains the substance serotonin.

Urtica products are marketed and used worldwide. Some commercial preparations containing Urtica (herb, root and leaf) are listed in Martindale 30 (1993), 32 (1999) and 33 (2002) (see Tables 7.1, 7.2, 7.3). Table 7.4 shows 'Products containing *Urtica* species in the German market'. *Urtica* is listed by Food and Drug Administration (FDA) as a herb of undefined safety. There are many the *Urtica* products sold over the counter (OTC) all over the world. Trade names in USA: Nettles capsules[R], Nettles Liquid extract[R] (Pittler, 2001).

Warning: The data and the information given in this chapter are not for the use as self medication without consulting with physicians.

References

Bisset, N. G. (1994) *Herbal Drugs and Phytopharmaceuticals*. Translated from Prof Dr Max Wichtl's original German edition. CRC Press, London, UK, pp. 502–9.
BHP (1983;1990/1) (*British Herbal Pharmacopeia*), British Herbal Medicine Association, Bournemouth, UK.
DAB 10 (1991) *Deutsches Arzneibuch* (German Pharmacopeia) 10th edn. Deutscher Apotheker Verlag, Stuttgart, Germany.
DAC (1986, 1987) *German Commission E Monographs*, American Botanical Council, Austin, Texas (USA).

ESCOP (1996, 1997) *European Scientific Cooperative on Phytotherapy,* Fascicule 2 (1996) Fascicule 4 (1997) revised in 2001. Monographs on the medicinal uses of plant drugs. Centre Complementary Health Studies, University of Exeter, UK.

Martindale (28th edn (1982), 30th edn (1993), 32nd edn (1999), 33rd edn (2002)) *The Extra Pharmacopeia,* eds J. E. F. Reynolds and S. C. Sweetman, Pharmaceutical Press, London, UK and Chicago.

Mills, S. Y. (2000) Personal communication. Use of nettle therapies in UK herbal medicine.

Mittman, P. (1990) Randomised double-blind study of freeze-dried *Urtica dioica* in the treatment of allergic rhinitis. *Planta Medica*, **56**(1): 44–7.

Newall, C. A., Anderson, L. A. and Phillipson, J. D. (1996) *Herbal Medicines – A Guide for Health-Care Professionals.* The Pharmaceutical Press, London, UK, pp. 201–2.

Pittler, M. H. (2001) *Nettle*, eds, Ernst, E., Pittler, M. H., Stevinsen, C. and White, A. R. *The Desktop Guide to Complementary and Alternative Medicine*, Mosby, Edinburgh, pp. 138–9.

Randall, C. F. (2001) Treatment of musculoskeletal pain with the sting of the stinging nettle: *Urtica dioica*. MD Thesis. University of Plymouth, Plymouth, UK.

Randall, C. F., Randall, H., Hutton, C., Sanders, H. and Dobbs, F. (2000) Randomized controlled trial of nettle sting for treatment of base-of-thumb pain. *Journal of Royal Society of Medicine*, **93**, 305–9.

Rote liste (2002) *Arzneimittel verzeichnis für Deutschland*.

Weiss, R. F. and Fintelmann, V. (2000) *Herbal Medicine*, 2nd edn (revised and expanded), Georg Thieme Verlag, New York, Stuttgart.

WHO (2002) *WHO Monographs on Selected Medicinal Plants*, World Health Organization, Geneva, Vol. 2, pp. 329–41.

Index